Test Yourself

Trigonometry

Mark N. Weinfeld, M.S.
MATHWORKS
New York, NY

Contributing Editors

Carl E. Langenhop, Ph.D.
Emeritus Professor of Mathematics
Southern Illinois University - Carbondale
Carbondale, IL

Tony Julianelle, Ph.D.
Department of Mathematics
University of Vermont
Burlington, VT

Charles M. Jones, M.S.
Department of Mathematics
East Texas State University
Commerce, TX

NTC LearningWorks
a division of NTC Publishing Group
Lincolnwood, Illinois

Library of Congress Cataloging-in-Publication Data
is available from the Library of Congress.

A *Test Yourself Books, Inc.* Project

Published by NTC Publishing Group
© 1996 NTC Publishing Group, 4255 West Touhy Avenue
Lincolnwood (Chicago), Illinois 60646-1975 U.S.A.

6 7 8 9 ML 0 9 8 7 6 5 4 3 2 1

Contents

Preface

Practice and more practice is the only way to truly master a subject area in mathematics. This is particularly true of trigonometry, which requires familiarity with a large number of definitions, formulas, and identities.

If you are currently taking a trigonometry course, this book is intended to serve as a supplement to whatever textbook you may be using. On the other hand, if your goal is to review topics you may have forgotten, this book will stand on its own. It contains a large number of problems of all types, along with detailed solutions. As you work through the problems in this book, you will be improving your mastery of trigonometry.

The format of each chapter is identical. Each begins with a short review called "Brief Yourself," which presents a concise summary of the key facts and formulas needed to solve the problems in that chapter. Then comes the "Test Yourself" section, which contains problems for you to solve. Solutions are given in the "Check Yourself" section. Finally, the "Grade Yourself" section helps you analyze your results and makes recommendations for further study.

There are several different ways in which this book can be used. If you are studying trigonometry from a standard textbook, you can use this book to test your comprehension of each chapter as you complete it. Simply turn to the chapter in this book that covers the material you have just studied. Then, begin by reading "Brief Yourself" to get a brief review. Proceed to work all of the problems in the chapter and check your answers. The "Grade Yourself" section will help you determine if you know the material well enough to move on, or if there are topics you must restudy.

Alternatively, if there is a specific type of problem you are struggling with, you can look up problems of that type in this book, and practice until you master them. Study the solutions carefully, as they will help you to understand how to think about such problems as you encounter them in the future.

The book is also useful if you want to relearn trigonometry and need a way of assessing exactly what you remember and what you need to relearn. Simply work through the book and use the "Grade Yourself" section to guide you as to your level of mastery of each topic.

This book covers all of the topics typically covered in a first-year trigonometry course and will be of value to anyone learning trigonometry for the first time, as well as to anyone who just needs to brush up on certain topics.

This book is for Brian, in the hope that someday he will dedicate one of his books to me.

Mark Weinfeld

How to Use this Book

This "Test Yourself" book is part of a unique series designed to help you improve your test scores on almost any type of examination you will face. Too often, you will study for a test—quiz, midterm, or final—and come away with a score that is lower than anticipated. Why? Because there is no way for you to really know how much you understand a topic until you've taken a test. The *purpose* of the test, after all, is to test your complete understanding of the material.

The "Test Yourself" series offers you a way to improve your scores and to actually test your knowledge at the time you use this book. Consider each chapter a diagnostic pretest in a specific topic. Answer the questions, check your answers, and then give yourself a grade. Then, and only then, will you know where your strengths and, more importantly, weaknesses are. Once these areas are identified, you can strategically focus your study on those topics that need additional work.

Each book in this series presents a specific subject in an organized manner, and although each "Test Yourself" chapter may not correspond to exactly the same chapter in your textbook, you should have little difficulty in locating the specific topic you are studying. Written by educators in the field, each book is designed to correspond, as much as possible, to the leading textbooks. This means that you can feel confident in using this book, and that regardless of your textbook, professor, or school, you will be much better prepared for anything you will encounter on your test.

Each chapter has four parts:

 Brief Yourself. All chapters contain a brief overview of the topic that is intended to give you a more thorough understanding of the material with which you need to be familiar. Sometimes this information is presented at the beginning of the chapter, and sometimes it flows throughout the chapter, to review your understanding of various *units* within the chapter.

 Test Yourself. Each chapter covers a specific topic corresponding to one that you will find in your textbook. Answer the questions, either on a separate page or directly in the book, if there is room.

Check Yourself. Check your answers. Every question is fully answered and explained. These answers will be the key to your increased understanding. If you answered the question incorrectly, read the explanations to *learn* and *understand* the material. You will note that at the end of every answer you will be referred to a specific subtopic within that chapter, so you can focus your studying and prepare more efficiently.

 Grade Yourself. At the end of each chapter is a self-diagnostic key. By indicating on this form the numbers of those questions you answered incorrectly, you will have a clear picture of your weak areas.

There are no secrets to test success. Only good preparation can guarantee higher grades. By utilizing this "Test Yourself" book, you will have a better chance of improving your scores and understanding the subject more fully.

Angle Measurement

Brief Yourself

Trigonometry can be subdivided into two parts: triangle trigonometry and analytic trigonometry. Triangle trigonometry is the study of angles and triangles. The notion of angles and their measurement is fundamental to this part of trigonometry.

An angle is formed by rotating a ray about its endpoint, which is called the vertex of the angle. The beginning position of the ray is called the initial side of the angle and the final position is called the terminal side of the angle. If the ray rotates in a counterclockwise direction a positive angle is formed; the angle is considered negative if the ray opens in the clockwise direction. An angle that is positioned with its vertex at the origin of a rectangular coordinate system with the initial side lying along the positive x-axis is said to be in standard position. Two angles in standard position whose terminal sides coincide are said to be coterminal angles.

Angles can be measured in degrees or in radians. An angle formed by one full revolution in the counterclockwise direction has a measure of 360 degrees, denoted 360°. The radian measure of this angle is 2π radians. Thus, one degree equals $(2\pi/360)$ radians and one radian equals $(360/2\pi)$ degrees.

A right angle measures 90° or $\pi/2$ radians. Any triangle containing a right angle is called a right triangle. An angle whose measure is between 0° and 90° is called an acute angle. A triangle in which all the angles are acute is called an acute triangle. An angle whose measure is greater than 90° but less than 180° is called an obtuse angle. An obtuse triangle is a triangle containing an obtuse angle.

Test Yourself

Sketch each of the following angles in standard position, labeling the initial side and the terminal side.

1. 60°

2. 90°

3. −30°

4. 45°

5. 120°

6. -90°

7. 270°

8. 180°

9. The terminal side of an angle in standard position lies in the first quadrant. What are the possible values of degree measure for this angle?

10. The terminal side of an angle in standard position lies in the third quadrant. What are the possible values of degree measure for this angle?

11. An angle of −77° is in standard position. In what quadrant is its terminal side?

12. An angle of −181° is in standard position. In what quadrant is its terminal side?

For each of the following angles, find a positive angle of less than one revolution that is coterminal with the given angle.

13. 400°

14. -800°

15. 900°

16. 5410°

17. Define what it means for an angle to be in standard position.

18. What are coterminal angles?

19. How many degrees equal one revolution? One quarter revolution? One half revolution?

20. How many radians equal one revolution? One quarter revolution? One half revolution?

Change the following degree measures to radian measure.

21. 30°

22. 45°

23. 90°

24. −150°

25. 120°

26. 145°

27. 180°

28. 75°

Change the following radian measures to degree measure.

29. $\pi/10$

30. 1

31. $3\pi/2$

32. 1.6

33. $-2\pi/3$

34. $5\pi/4$

35. −2.2

36. $4\pi/5$

37. What is a quadrantal angle?

38. How do you classify an angle by quadrant? Does an angle have to be in standard position to be classified as to quadrant?

Classify each of the following angles by quadrant or identify it as a quadrantal angle.

39. 150°

40. −370°

41. 493°

42. 1440°

For each of the following angles, find a positive angle of less than one revolution that is coterminal with the given angle.

43. 5π

44. $19\pi/4$

45. −5

46. 13.279

✓ Check Yourself

1.

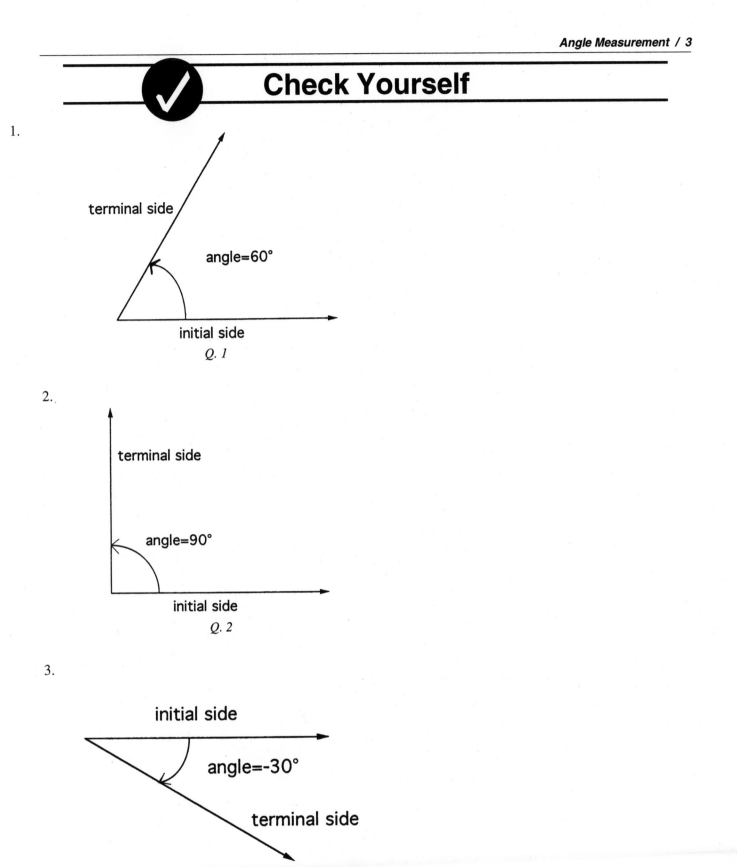

terminal side

angle=60°

initial side

Q. 1

2.

terminal side

angle=90°

initial side

Q. 2

3.

initial side

angle=-30°

terminal side

Q. 3

4.

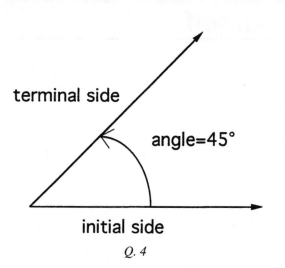

Q. 4

5.

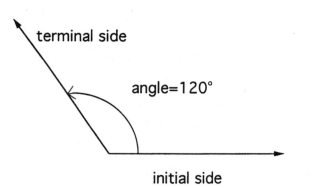

Q. 5

6.

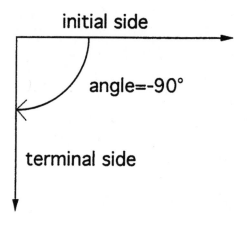

Q. 6

7.

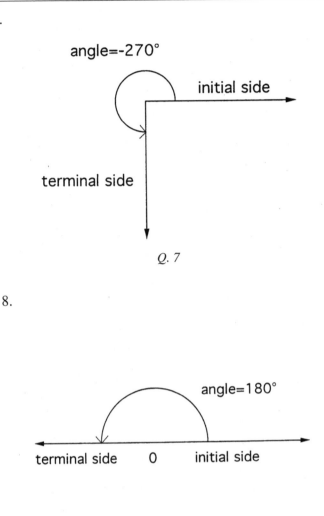

Q. 7

8.

Q. 8

9. The degree measure of a first quadrant angle is either between 0 and 90° or else equals a number obtained by adding any positive or negative integer multiple of 360° to an angle between 0 and 90°; that is, $\theta° + n360°$ where $0 < \theta < 90$ and $n = 0, \pm1, \pm2,...$ **(Degree measure)**

10. The degree measure of a third quadrant angle is either between 180° and 270°, or else equals a number obtained by adding any positive or negative integer multiple of 360° to an angle between 180° and 270°; that is, $\theta° + n360°$ where $180 < \theta < 270$ and $n = 0, \pm1, \pm2,...$ **(Degree measure)**

11. An angle of −77° corresponds to a *clockwise* rotation (since it is negative). Since the angle is between 0° and −90°, its terminal side is in quadrant IV. **(Degree measure)**

12. An angle of −181° corresponds to slightly more than one half revolution in a clockwise direction; i.e., its terminal side is in quadrant II. **(Degree measure)**

13. 400° = 360° + 40°. A 40° angle is coterminal with a 400° angle. **(Degree measure)**

14. −800° = −1080° + 280°. A 280° angle is coterminal with an −800° angle. **(Degree measure)**

15. 900°= 720° + 180°. An angle of 180° is coterminal with an angle of 900°. **(Degree measure)**

16. 5410° = 15(360°) + 10° A 10° angle is coterminal with an angle of 5410°. (**Degree measure**)

17. An angle is in standard position relative to a rectangular coordinate system if its vertex is located at the origin and its initial side lies along the positive x-axis. (**Degree measure**)

18. Two angles in standard position are coterminal if they have the same terminal side. There are infinitely many coterminal angles for any angle. (**Degree measure**)

19. One revolution equals 360 degrees. One quarter revolution equals 90 degrees, and there are 180 degrees in a half revolution. (**Degree measure**)

20. 2π radians equal one revolution. One quarter revolution equals $\pi/2$ radians, and there are π radians in a half revolution. (**Radian measure**)

21. $360° = 2\pi$ rad. Therefore $1° = \pi/180$ rad. and $30° = 30(\pi/180) = \pi/6$ radians (**Changing degrees to radians**)

22. $45° = 45(\pi/180) = \pi/4$ radians (**Changing degrees to radians**)

23. $90° = 90(\pi/180) = \pi/2$ radians (**Changing degrees to radians**)

24. $-150° = -150(\pi/180) = -5\pi/6$ radians (**Changing degrees to radians**)

25. $120° = 120(\pi/180) = 2\pi/3$ radians (**Changing degrees to radians**)

26. $145° = 145(\pi/180) = 27\pi/36$ radians = $3\pi/4$ radians (**Changing degrees to radians**)

27. $180° = 180(\pi/180) = \pi$ radians (**Changing degrees to radians**)

28. $75° = 75(\pi/180) = 5\pi/12$ radians (**Changing degrees to radians**)

29. 2π rad = 360°. Therefore 1 rad. = $180/\pi$ degrees and $\pi/10$ rad = $\pi/10$ $(180/\pi)$ = 18° (**Changing radians to degrees**)

30. 1 rad = $180/\pi$ degrees $\cong$ 57.2958 degrees (**Changing radians to degrees**)

31. $3\pi/2$ rad = $3\pi/2$ $(180/\pi$ degrees) = 270 degrees = 270° (**Changing radians to degrees**)

32. 1.6 rad = 1.6 $(180/\pi$ degrees) $\cong$ 91.6732 degrees = 91.6732° (**Changing radians to degrees**)

33. $-2\pi/3$ rad = $-2\pi/3(180/\pi$ degrees) = -120 degrees = $-120°$ (**Changing radians to degrees**)

34. $5\pi/4$ rad = $5\pi/4(180/\pi$ degrees) = 225 degrees = 225° (**Changing radians to degrees**)

35. -2.2 rad = -2.2 $(180/\pi$ degrees) $\cong$ -126.0507 degrees = $-126.0507°$ (**Changing radians to degrees**)

36. $4\pi/5$ rad = $4\pi/5$ $(180/\pi$ degrees) = 144 degrees = 144° (**Changing radians to degrees**)

37. A quadrantal angle is an angle in standard position whose terminal side lies on the positive or negative x- or y-axis. (**Angles and quadrants**)

38. An angle in standard position is classified by quadrant according to which quadrant contains the terminal side. Thus, an angle between 0° and 90° is a quadrant I angle. The angle must be in standard position to be classified by quadrant. (**Angles and quadrants**)

39. 150° is a quadrant II angle. (**Degree measure**)

40. −370° = −360° − 10° is a quadrant IV angle. (**Degree measure**)

41. 493° = 360° + 133° is a quadrant II angle. (**Degree measure**)

42. 1440° = 4(360°) is a quadrantal angle. (**Degree measure**)

43. 5π rad = $2(2\pi)$ + π rad. is coterminal with the angle π rad. (**Radian measure**)

44. $19\pi/4$ rad. = 4π + $3\pi/4$ rad. is coterminal with $3\pi/4$. (**Radian measure**)

45. −5 rad. is coterminal with $(2\pi - 5)$ rad. = 1.2832 rad. (**Radian measure**)

46. 13.279 rad = 4π + .7126 rad is coterminal with 0.7126 rad. (**Radian measure**)

Grade Yourself

Circle the numbers of the questions you missed, then fill in the total incorrect for each topic. If you answered more than three questions incorrectly, you need to focus on that topic. (If a topic has less than three questions and you had at least one wrong, we suggest you study that topic also. Read your textbook, a review book, or ask your teacher for help.)

Subject: Angle Measurement

Topic	Question Numbers	Number Incorrect
Degree measure	1, 2, 3, 4, 5, 6, 7, 8, 9, 10, 11, 12, 13, 14, 15, 16, 17, 18, 19, 39, 40, 41, 42	
Radian measure	20, 43, 44, 45, 46	
Changing degrees to radians	21, 22, 23, 24, 25, 26, 27, 28	
Changing radians to degrees	29, 30, 31, 32, 33, 34, 35, 36	
Angles and quadrants	37, 38	

Trigonometric Functions

Brief Yourself

If point P on the terminal side of angle A in standard position is located a positive distance r from the origin and has coordinates (x,y), then the six trigonometric functions of A are defined as follows:

sine of A = sin A = y/r   cosecant of A = csc A = r/y (if y ≠ 0)

cosine of A = cos A = x/r   secant of A = sec A = r/x (if x ≠ 0)

tangent of A = tan A = y/x   (if x ≠ 0)   cotangent of A = cot A = x/y (if y ≠ 0)

where $r = \sqrt{x^2 + y^2} > 0$ equals the distance of P from the origin.

Note that $\dfrac{1}{\sin A} = $ csc A, $\dfrac{1}{\cos A} = $ sec A,   and $\dfrac{1}{\tan A} = $ cot A. For this reason, cosecant, secant, and cotangent are referred to as reciprocal functions.

The coordinates x and y may be positive or negative depending on the quadrant in which the terminal side of A lies. Thus, the sine and cosecant are positive when A is a quadrant I or II angle; the cosine and secant are positive when A is an angle in quadrant I or IV; the tangent and cotangent are positive if A is an angle in quadrant I or III.

The trigonometric functions can be evaluated by means of a calculator when the measure of A is known. Refer to the operator's manual to find how to specify degree or radian measure on your calculator. Most calculators only contain keys for sine, cosine and tangent. The reciprocal functions can be evaluated using the reciprocal key, 1/x.

The reference angle A′ of an angle A in standard position is the smaller of the two positive angles formed by the terminal side of A and the x-axis. Since A′ is always between 0° and 90°, the six trigonometric functions of A′ are always positive (or undefined if A′ equals 0° or 90°). The trigonometric functions of A always have the same numerical value as those of A′, but the trigonometric functions of A may be negative depending on the quadrant in which the terminal side of A is found.

The formulas f(x) = sin x and f(x) = cos x (x measured in radians) each produce a well-defined value for all values of x, hence each formula defines a function. The sine and cosine functions are defined for all values of x. The assumed values range from −1 to +1. There are certain values of x for which the functions tan x, cot x, csc x and sec x are undefined.

Any function f(x) with the property that f(−x) = f(x) is called an even function. The functions f(x) = cos x and f(x) = sec x are even functions. Any function f(x) with the property that −f(−x) = f(x) is called an odd function. The four remaining trig functions are odd functions.

The graphs of the functions f(x) = sin x and f(x) = cos x oscillate between the values +1 and −1. We say the amplitude of these functions equals 1. The graphs of the sine and cosine functions repeat their oscillatory cycle in intervals of length 2π; the functions are said to be periodic with period 2π. More generally, the functions f(x) = A sin Px and f(x) = A cos Px have amplitude equal to |A| and are periodic with period 2π/|P|.

Test Yourself

In problems 1 to 8, a point on the terminal side of angle *a* is given. Assuming *a* is in standard position, evaluate the six trigonometric functions of *a*.

1. (3,4)

2. (5,12)

3. ($\sqrt{3}$,1)

4. (2,0)

5. (1/3,1/4)

6. (1/3,−3/4)

7. ($\sqrt{7}$,−3)

8. (0,4)

Without using a calculator, evaluate all six trigonometric functions for each of the following angles or state that a function is not defined:

9. 45°

10. π/6

11. −90°

12. 3π/4

13. −π/3

14. 180°

15. 330°

16. 5π/2

Using a calculator, find the value of the following functions to four-place accuracy:

17. sin 25°

18. cos 18°

19. tan 35°

20. sec 37°

21. cot 56°

22. csc 80°

23. sin 1.1

24. cos .95

25. tan π/7

26. cot π/9

27. sec 1

28. csc 1.2

Identify all possible quadrants in which the following statements are true for angle a in standard position:

29. $\sin a > 0$

30. $\sin a < 0$

31. $\sin a > 0$ and $\cos a < 0$

32. $\cos a > 0$

33. $\tan a < 0$

34. $\tan a > 0$ and $\sec a > 0$

35. $\sec a < 0$

36. $\sin a < 0$ and $\cos a < 0$.

Without using a calculator, evaluate the six trigonometric functions for the angle a which satisfies:

37. $\cos a = -0.5$, a in quadrant II

38. $\tan a = 6$, a in quadrant III

39. $\sin a = -5/13$, a in quadrant IV

40. $\sec a = 2$, a in quadrant I

41. $\cos a = 12/13$, $\tan a < 0$

42. $\cot a = 15/8$, $\cos a < 0$

43. $\sin a = -3/5$, $\cos a < 0$

44. $\tan a = 2$, $\cos a > 0$

Find the reference angle for the following angles in standard position:

45. 330°

46. $3\pi/2$

47. 160°

48. $5\pi/12$

49. $5\pi/3$

50. 22°

51. 226°

52. $5\pi/6$

53. Define a reference angle.

Express the indicated functions in terms of the following reference angles:

54. $\tan 336°$

55. $\cos 7\pi/12$

56. $\sin 355°$

57. $\cos 265°$

58. $\sec 2\pi/5$

59. $\csc 148°$

60. $\sec 320°$

61. $\tan 7\pi/12$

62. $\sin 22\pi/29$

Using a calculator, find all angles a, $0 < a < 90°$, having the given function values. Give answers in degrees rounded to two places.

63. $\cos a = 0.5$

64. $\tan a = 1.800$

65. $\sin a = .3101$

66. $\cos a = 0.750$

67. $\tan a = 6.800$

68. $\sin a = 4.3$

Using a calculator when necessary, find all the angles of less than one revolution having the given function values. When using a calculator, give answers in radians rounded to four places.

69. $\cos a = -0.5$

70. $\sin a = -.7071$

71. $\tan a = -7.1153$

72. cos a = 0.4076

73. sin a = .7547

74. cos a = −1.115

Use the periodicity and odd-even properties of the trigonometric functions to find the exact values of the following functions:

75. tan 21π

76. cos 405°

77. sin 420°

78. cos 390°

79. sec 450°

80. csc 540°

81. tan 25π/6

82. csc 9π/2

83. sec 9π/4

84. sin(−π/6)

85. cos(−π/6)

86. tan(−π/4)

87. cos(−3π/2)

88. csc(−π/3)

89. sec(−135°)

For each of the following functions, give all values of x where f(x) is not defined and all values where f(x) = 0. Also, give the range of values assumed by f(x) as x varies over all allowed values.

90. f(x) = sin x

91. f(x) = cos x

92. f(x) = tan x

93. f(x) = csc x

94. f(x) = sec x

95. f(x) = cot x

Find the amplitude and the period and sketch the graph for the following functions:

96. y(x) = 4 sin 2x

97. y(x) = 2 sin(x/3)

98. y(x) = 3 cos 3x

99. y(x) = − sin πx

100. y(x) = −2 sin(πx/2)

Find the amplitude, the period, and an equation for the functions having the following graphs:

101.

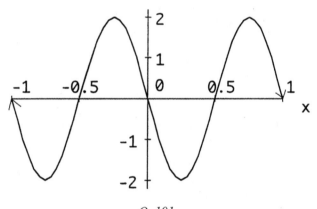

Q. 101

102.

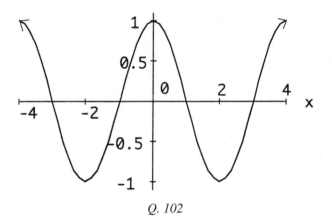

Q. 102

103.

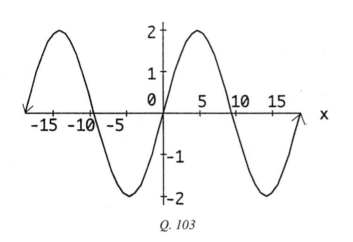

Q. 103

104.

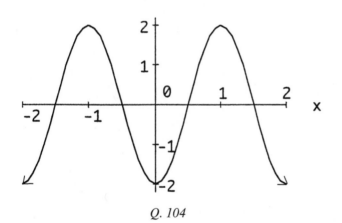

Q. 104

105.

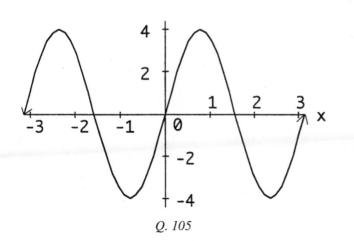

Q. 105

✓ Check Yourself

1. sin a = 4/5   cos a = 3/5   tan a = 4/3
    csc a = 5/4   sec a = 5/3   cot a = 3/4.

2. sin a = 12/13   cos a = 5/13   tan a = 12/5
    csc a = 13/12   sec a = 13/5   cot a = 5/12.

3. sin a = −1/2   cos a = −$\sqrt{3}$/2   tan a = 1/$\sqrt{3}$
    csc a = −2   sec a = −2/$\sqrt{3}$ cot a = $\sqrt{3}$

4. sin a = 0   cos a = 1   tan a = 0
    csc a = undefined sec a = 1   cot a = undefined

5. sin a = 3/5   cos a = −4/5   tan a = −3/4
    csc a = 5/3   sec a = −5/4   cot a = −4/3

6. sin a = −9/$\sqrt{97}$ cos a = 4/$\sqrt{97}$ tan a = −9/4
    csc a = −$\sqrt{97}$/9   sec a = $\sqrt{97}$/4   cot a = −4/9

7. sin a = −3/4   cos a = $\sqrt{7}$/4   tan a = −3/$\sqrt{7}$
    csc a = −4/3   sec a = 4/$\sqrt{7}$ cot a = −$\sqrt{7}$/3

8. sin a = 1   cos a = 0   tan a = undefined
    csc a = 1   sec a = undefined cot a = 0

    (Definition of circular functions)

9. sin 45° = 1/$\sqrt{2}$ cos 45° = 1$\sqrt{2}$ tan 45° = 1
    csc 45° = $\sqrt{2}$ sec 45° = $\sqrt{2}$ cot 45° = 1

10. sin π/6 = 1/2   cos π/6 = $\sqrt{3}$/2   tan π/6 = 1/$\sqrt{3}$
    csc π/6 = 2   sec π/6 = 2/$\sqrt{3}$ cot π/6 = $\sqrt{3}$

11. sin (−90°) = −1   cos (−90°) = 0   tan (−90°) = undefined
    csc (−90°) = −1   sec (−90°) = undefined   cot (−90°) = 0

12. sin (−3π/4) = −1/$\sqrt{2}$ cos (−3π/4) = −1/$\sqrt{2}$ tan (−3π/4) = 1
    csc (−3π/4) = −$\sqrt{2}$ sec (−3π/4) = −$\sqrt{2}$ cot (−3π/4) = 1

13. sin(−π/3) = −$\sqrt{3}$/2   cos(−π/3) = 1/2   tan(−π/3) = −$\sqrt{3}$
    csc(−π/3) = −2/$\sqrt{3}$ sec(−π/3) = 2   cot(−π/3) = −1/$\sqrt{3}$

14. sin 180° = 0   cos 180° = −1   tan 180° = 0
    csc 180° = undefined   sec 180° = −1   cot 180° = undefined

15. sin 330° = −1/2   cos 330° = $\sqrt{3}$/2   tan 330° = −1/$\sqrt{3}$
    csc 330° = −2   sec 330° = 2/$\sqrt{3}$ cot 330° = −$\sqrt{3}$

16. $\sin 5\pi/2 = 1$ $\cos 5\pi/2 = 0$ $\tan 5\pi/2 = $ undefined

      $\csc 5\pi/2 = 1$ $\sec 5\pi/2 = $ undefined $\cot 3\pi/4 = 0$

(Evaluating trigonometric functions)

17. $\sin 25° = .4226$

18. $\cos 18° = .9511$

19. $\tan 35° = .7002$

20. $\sec 37° = 1.2521$

21. $\cot 56° = .6745$

22. $\csc 80° = 1.0154$

23. $\sin 1.1 = .8912$

24. $\cos .95 = .5817$

25. $\tan \pi/7 = .4816$

26. $\cot \pi/9 = 2.7475$

27. $\sec 1 = 1.8508$

28. $\csc 1.2 = 1.0729$

(Evaluating trigonometric functions)

29. $\sin a$ is positive in quadrants I and II.

30. $\sin a$ is negative in quadrants III and IV.

31. $\sin a$ is positive in quadrants I and II and $\cos a$ is negative in quadrants II and III. Thus, $\sin a$ is positive and $\cos a$ is negative in quadrant II (only).

32. $\cos a$ is positive in quadrants I and IV.

33. $\tan a$ is negative in quadrants where $\sin a$ and $\cos a$ have opposite signs. This happens in quadrants II and IV.

34. $\tan a$ is positive in quadrants I and III, while $\sec a$ is positive in quadrants I and IV. Thus, $\tan a$ and $\sec a$ are both positive in quadrant I (only).

35. $\sec a$ is negative in quadrants II and III.

36. $\sin a$ is negative in quadrants III and IV, while $\cos a$ is negative in quadrants II and III. Thus, they are both negative in quadrant III.

(Properties of trigonometric functions)

37. Since a is a quadrant II angle with cos $a = -1/2$, the point $(-1, \sqrt{3})$ must lie on the terminal side of the angle. Then

sin $a = \sqrt{3}/2$,	cos $a = -1/2$,	tan $a = -\sqrt{3}$
csc $a = 2/\sqrt{3}$,	sec $a = -2$,	cot $a = -1/\sqrt{3}$

38. Since a is a quadrant III angle with tan $a = 6$, the point $(-1,-6)$ must lie on the terminal side of the angle. Then

sin $a = -6/\sqrt{37}$,	cos $a = -1/\sqrt{37}$,	tan $a = 6$
csc $a = -\sqrt{37}/6$,	sec $a = -\sqrt{37}$,	cot $a = 1/6$

39. Here a is a quadrant IV angle with sin $a = -5/13$, hence the point $(12,-5)$ must lie on the terminal side of the angle. Then

sin $a = -5/13$,	cos $a = 12/13$,	tan $a = -5/12$
csc $a = -13/5$,	sec $a = 13/12$,	cot $a = -12/5$

40. If a is a quadrant I angle with sec $a = 2$, then the point $(1,\sqrt{3})$ must lie on the terminal side of the angle. Then

sin $a = \sqrt{3}/2$,	cos $a = 1/2$,	tan $a = \sqrt{3}$
csc $a = 2/\sqrt{3}$,	sec $a = 2$,	cot $a = 1/\sqrt{3}$

41. Since cos $a = 12/13$ and tan a is negative, a must be a quadrant IV angle having the point $(12, -5)$ on its terminal side. Then

sin $a = -5/13$,	cos $a = 12/13$,	tan $a = -5/12$
csc $a = -13/5$,	sec $a = 13/12$,	cot $a = -12/5$

42. Since cot $a = 15/8$ and cos a is negative, a must be a quadrant III angle having the point $(-15, -8)$ on its terminal side. Then

sin $a = -8/17$,	cos $a = -15/17$,	tan $a = 8/15$
csc $a = -17/8$,	sec $a = -17/15$,	cot $a = 15/8$

43. Since sin a and cos a are both negative, a is a quadrant III angle. Since sin $a = -3/5$, the point $(-4,-3)$ is a point on the terminal side of the angle. Then

sin $a = -3/5$,	cos $a = -4/5$,	tan $a = 3/4$
csc $a = -5/3$,	sec $a = -5/4$,	cot $a = 4/3$

44. Since tan a and cos a are both positive, a is a quadrant I angle. Since tan $a = 2$, the point $(1,2)$ is a point on the terminal side of the angle. Then

sin $a = 2/\sqrt{5}$,	cos $a = 1/\sqrt{5}$,	tan $a = 2$
csc $a = \sqrt{5}/2$,	sec $a = \sqrt{5}$,	cot $a = 1/2$

(Properties of trigonometric functions)

45. The reference angle for $330°$ is $360°-330°= 30°$.

46. The reference angle for $3\pi/2$ radians is $2\pi - 3\pi/2 = \pi/2$.

47. The reference angle for $-160°$ is $180°-160°= 20°$.

48. The reference angle for the quadrant I angle $5\pi/12$ radians is $5\pi/12$.

49. The reference angle for $5\pi/3$ radians is $2\pi - 5\pi/3 = \pi/3$.

50. The reference angle for the acute angle $22°$ is $22°$.

51. The reference angle for $226°$ is $226° - 180° = 46°$.

52. The reference angle for $-5\pi/6$ radians is $\pi - 5\pi/6 = \pi/6$.

53. The reference angle a' for angle a in standard position is the smaller of the two positive angles formed by the terminal side of a and the x–axis.

(Reference angle)

54. The reference angle for the quadrant IV angle $336°$ is $360° - 336° = 24°$. Hence, $\tan 336° = \pm \tan 24°$. Since the tangent of a quadrant IV angle is negative, $\tan 336° = -\tan 24°$.

55. The reference angle for the quadrant II angle $7\pi/12$ radians is $\pi - 7\pi/12 = 5\pi/12$. Hence, $\cos 7\pi/12 = \pm \cos 5\pi/12$. The cosine of a quadrant II angle is negative, and therefore $\cos 7\pi/12 = -\cos 5\pi/12$.

56. The reference angle for the quadrant IV angle $355°$ is $360° - 355° = 5°$. Since $\sin 355° = \pm \sin 5°$, and the sine of a quadrant IV angle is negative, $\sin 355° = -\sin 5°$.

57. The reference angle for the quadrant III angle $265°$ is $265° - 180° = 85°$. Since $\cos 265° = \pm \cos 85°$, and the cosine of a quadrant III angle is negative, $\cos 265° = -\cos 85°$.

58. $2\pi/5$ is a first quadrant angle. Therefore, $\sec 2\pi/5$ is already expressed in terms of the reference angle.

59. The reference angle for the quadrant II angle $a = 148°$ is $180° - 148° = 32°$. The cosecant of a quadrant II angle is positive, hence $\csc 148° = \csc 32°$.

60. The reference angle for the quadrant IV angle $320°$ is $360° - 320° = 40°$. Since $\sec 320° = \pm \sec 40°$, and the secant of a quadrant IV angle is positive, $\sec 320° = \sec 40°$.

61. The reference angle for the quadrant II angle $7\pi/12$ radians is $\pi - 7\pi/12 = 5\pi/12$. The tangent is negative in quadrant II, and thus $\tan 7\pi/12 = -\tan 5\pi/12$.

62. $a = 22\pi/29$ is a quadrant II angle with reference angle $a' = 7\pi/29$. The sine of a quadrant II angle is positive, thus $\sin 22\pi/29 = \sin 7\pi/29$.

(Evaluating trigonometric functions)

63. $a = 60°$ is the only angle having $\cos a = 0.5$, $0 < a < 90°$.

64. $a = 60.95°$ is the only angle having $\tan a = 1.800$, $0 < a < 90°$.

65. $a = 18.07°$ is the only angle having $\sin a = 0.3101$, $0 < a < 90°$.

66. $a = 41.41°$ is the only angle having $\cos a = 0.750$, $0 < a < 90°$.

67. $a = 81.63°$ is the only angle having $\tan a = 6.800$, $0 < a < 90°$.

68. Since $-1 \le \sin a \le 1$ for all angles a, this equation has no solution.

69. Since the cosine is negative only in quadrants II and III, a must be a quadrant II or quadrant III angle with reference angle a' such that $\cos a' = 0.5$. Then $a' = \pi/3$, and $a = \pi - \pi/3 = 2\pi/3$ is the quadrant II solution, while $a = \pi + \pi/3 = 4\pi/3$ is the solution in quadrant III.

70. The sine is negative only in quadrants III and IV, so a must be a quadrant III or quadrant IV angle with reference angle a' such that $\sin a' = 0.7071$. Then $a' = \pi/4$ and $a = \pi + \pi/4 = 5\pi/4$ is the quadrant III solution, while $a = 2\pi - \pi/4 = 7\pi/4$ is the solution in quadrant IV.

71. The tangent is negative in quadrants II and IV, so a must be a quadrant II or quadrant IV angle with reference angle a' such that $\tan a' = 7.1153$. Then $a' = 1.4312$ rad and $a = \pi - 1.4312 = 1.7104$ rad is the quadrant II solution, while $a = 2\pi - 1.4312 = 4.8520$ rad is the solution in quadrant IV.

72. The cosine is positive in the first and fourth quadrants, so a equals a quadrant I or quadrant IV angle with reference angle a' such that $\cos a' = .4076$. Then $a' = 1.1510$ rad and $a = 2\pi - 1.1510 = 5.1322$ rad is the quadrant IV solution, while $a = 1.1510$ rad is the solution in quadrant III.

73. The sine is positive in quadrants I and II, so a must be a quadrant I or quadrant II angle with reference angle a' such that $\sin a' = 0.7547$. Then $a' = .8552$ and $a = a' = .8552$ rad is the quadrant I solution, while $a = \pi - .8552 = 2.2864$ rad is the solution in quadrant II.

74. Since $-1 \le \cos a \le 1$ for all angles a, this equation has no solution.

(Evaluation of trigonometric functions)

75. $\tan 21\pi = \tan(10(2\pi) + \pi) = \tan \pi = 0$

76. $\cos 405° = \cos (360° + 45°) = \cos 45° = 1/\sqrt{2}$

77. $\sin 420° = \sin(360° + 60°) = \sin 60° = \sqrt{3}/2$

78. $\cos 390° = \cos(360° + 30°) = \cos 30° = \sqrt{3}/2$

79. $\sec 450° = \sec(360° + 90°) = \sec 90° = $ undefined

80. $\csc 540° = \csc(360° + 180°) = \csc 180° = $ undefined

81. $\tan 25\pi/6 = \tan(4\pi + \pi/6) = \tan \pi/6 = 1/\sqrt{3}$

82. $\csc 9\pi/2 = \csc(4\pi + \pi/2) = \csc \pi/2 = 1$

83. $\sec 9\pi/4 = \sec(2\pi + \pi/4) = \sec \pi/4 = \sqrt{2}$

84. $\sin(-\pi/6) = -\sin \pi/6 = -0.5$

85. $\cos(-\pi/6) = \cos \pi/6 = \sqrt{3}/2$

86. $\tan(-\pi/4) = -\tan \pi/4 = -1$

87. $\cos(-3\pi/2) = \cos 3\pi/2 = 0$

88. $\csc(-\pi/3) = -\csc \pi/3 = -2/\sqrt{3}$

89. $\sec(-135°) = \sec 135° = -\sec 45° = -\sqrt{2}$

(Properties of trigonometric functions)

90. $f(x) = \sin x$ is defined for all values of x. $\sin x = 0$ for x equal to any positive or negative integer multiple of π. Finally, $-1 \leq \sin x \leq 1$ for all angles x.

91. $f(x) = \cos x$ is defined for all values of x. $\cos x = 0$ for x equal to any odd integer multiple of $\pi/2$. Finally, $-1 \leq \cos x \leq 1$ for all angles x.

92. $f(x) = \tan x$ is undefined for x equal to any odd integer multiple of $\pi/2$. $\tan x = 0$ for a equal to any positive or negative integer multiple of π. $f(x) = \tan x$ ranges over all positive and negative values as x goes from $-\pi/2$ to $\pi/2$.

93. $f(x) = \csc x$ is undefined for all values of x such that $\sin x = 0$, i.e., for x equal to any positive or negative integer multiple of π. The cosecant function is never zero and, in fact, $|\csc x| \geq 1$ for all x. This means that $\csc x$ can assume only positive values greater than or equal to $+1$ or negative values less than or equal to -1 as x varies over its allowable values.

94. $f(x) = \sec x$ is undefined for all values of x such that $\cos x = 0$, i.e., for x equal to any odd integer multiple of $\pi/2$. The secant function is never zero and, in fact, $|\sec x| \geq 1$ for all x. This means that $\sec x$ can assume only positive values greater than or equal to $+1$ or negative values less than or equal to -1.

95. $f(x) = \cot x$ is undefined at each point where $\sin x = 0$; i.e., $\cot x$ is undefined for x equal to any positive or negative integer multiple of π. Also, $\cot x = 0$ for x equal to any odd integer multiple of $\pi/2$. Finally, $f(x) = \cot x$ ranges over all positive and negative values as x goes from 0 to π.

(Properties of trigonometric functions)

96. $y = 4 \sin 2x$ has amplitude equal to 4 and period equal to $P = 2\pi/2 = \pi$.

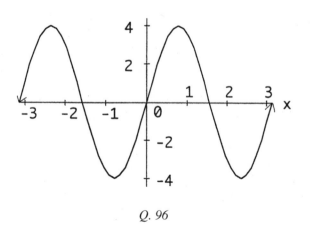

Q. 96

97. y = 2 sin(x/3) has amplitude equal to 2 and period P = 2π/(1/3) = 6π.

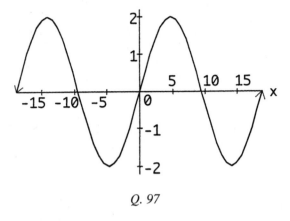

Q. 97

98. The amplitude of y = 3 cos 3x equals 3 and the period is P = 2π/3

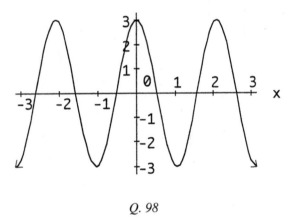

Q. 98

99. y = − sin πx has amplitude 1 and period equal to P = 2π/π = 2

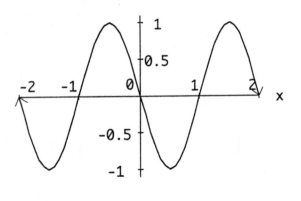

Q. 99

100. $y = -2 \sin(\pi x/2)$ has amplitude 2 and period $P = 2\pi/(\pi/2) = 4$

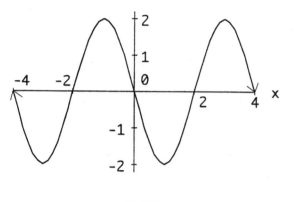

Q. 100

101. Since the graph oscillates between extremes of $y = 2$ and $y = -2$, the amplitude equals 2. One complete cycle of the graph is contained in the interval $[0,1]$ so the period equals $P = 1$. The sinusoidal shape of the graph indicates that the equation has the form $y = A \sin 2\pi x/P$ or $y = A \cos 2\pi x/P$. Since $y = 0$ at $x = 0$, it is the graph of a sine curve, and since y is negative between 0 and 0.5 and positive for x between .5 and 1, we conclude that $A = -2$ and $P = 1$. That is, $y = -2 \sin 2\pi x$.

102. This graph oscillates between 1 and -1, so the amplitude is equal to 1. One complete cycle of the graph is contained in the interval $[0,4]$ which implies that $P = 4$. The shape of the graph suggests that $y = A \sin 2\pi x/P$ or $y = A \cos 2\pi x/P$. Since the graph starts out at $y=1$ for $x=0$, it must be the latter. Then $A=1$ and $P=4$, so $y = \cos(\pi x/2)$.

103. Although it is difficult to tell the precise length of the period of this graph from the figure, one cycle is contained in the interval $[0,6\pi]$. The amplitude is clearly equal to 2 and the shape of the graph indicates this is the graph of a sine function. Thus,

$$y = 2 \sin(2\pi x/(6\pi)) = 2 \sin(x/3)$$

104. The amplitude of this graph equals 2, and one cycle is contained in the interval $[0,2]$. Since $y = -2$ at $x = 0$, this is the graph of a cosine function with $A = -2$ and $P = 2$. That is, $y = -2 \cos \pi x$.

105. The amplitude of this graph equals 4 and one cycle of the graph appears to be contained in the interval $[0,\pi]$, so that $P=\pi$. The sinusoidal shape of the graph and the fact that $y=0$ at $x=0$ indicate that this is the graph of a function of the form $y = A \sin 2\pi x/P$. Then $A = 4$ and $P = \pi$, so the equation must be $y = 4 \sin 2x$.

(Graphing trigonometric functions)

Grade Yourself

Circle the numbers of the questions you missed, then fill in the total incorrect for each topic. If you answered more than three questions incorrectly, you need to focus on that topic. (If a topic has less than three questions and you had at least one wrong, we suggest you study that topic also. Read your textbook, a review book, or ask your teacher for help.)

Subject: Trigonometric Functions

Topic	Question Numbers	Number Incorrect
Definition of circular functions	1, 2, 3, 4, 5, 6, 7, 8	
Evaluating trigonometric functions	9. 10, 11, 12, 13, 14, 15, 16, 17, 18, 19, 20, 21, 22, 23, 24, 25, 26, 27, 28, 54, 55, 56, 57, 58, 59, 60, 61, 62, 63, 64, 65, 66, 67, 68, 69, 70, 71, 72, 73, 74	
Properties of trigonometric functions	29, 30, 31, 32, 33, 34, 35, 36, 37, 38, 39, 40, 41, 42, 43, 44, 75, 76, 77, 78, 79, 80, 81, 82, 83, 84, 85, 86, 87, 88, 89, 90, 91, 92, 93, 94, 95	
Reference angle	45, 46, 47, 48, 49, 50, 51, 52, 53	
Graphing trigonometric functions	96, 97, 98, 99, 100, 101, 102, 103, 104, 105	

Graphs of Trigonometric Functions

Brief Yourself

The best way to understand the behavior of the trigonometric functions is by sketching their graphs.

Recall from chapter 2 that the graphs of the functions f(x) = sin x and f(x) = cos x have an amplitude of 1 and a period of length 2π radians = 360°. Also recall that functions of the form f(x) = A sin Px and f(x) = A cos Px have amplitude equal to $|A|$ and a period of $2\pi/|P|$ radians = $360/|P|$ degrees.

In general, functions of the form f(x) = A sin (Px + C) and f(x) = A cos (Px + C) have amplitude $|A|$, period $2\pi/|P|$ radians = $360/|P|$ degrees, and are "phase shifted" $|C/P|$ units to the left if C/P is positive, or $|C/P|$ units to the right if C/P is negative. Further, in graphs of the form f(x) = sin x + D and f(x) = cos x + D, the constant D lifts the entire curve up D units if D is positive and lowers the entire curve D units if D is negative.

While the graphs of f(x) = sin x and f(x) = cos x are continuous curves, the graphs of the other four trigonometric functions are discontinuous. The graph of f(x) = tan x is undefined at odd multiples of $\pi/2$ radians = 90°, corresponding to the values of x for which cos x = 0. As x gets closer to the values for which tan x is undefined, f(x) increases or decreases without limit; we say that the graph of f(x) = tan x has asymptotes wherever tan x is undefined. Asymptotes are indicated on graphs as dashed vertical lines.

Similarly, the graph of f(x) = cot x is undefined, and therefore has asymptotes, at integer multiples of π radians = 180°, corresponding to the values of x for which sin x = 0. Note that, unlike f(x) = sin x and f(x) = cos x, the graphs of both f(x) = tan x and f(x) = cot x are periodic with period π radians = 180°.

The cosecant function is the reciprocal of the sine function. Therefore, its graph is periodic with period 2π radians = 360°, and has asymptotes wherever sin x = 0. Similarly, the secant function is the reciprocal of the cosine function. Therefore, it too has period 2π radians = 360°, and has asymptotes wherever cos x = 0.

The comments regarding period changes, phase shifting, and lifting of the graphs of sin x and cos x also apply to the graphs of the other trigonometric functions.

Test Yourself

In problems 1 to 12, find the amplitude, period, phase shift, displacement, and sketch the graph for the following functions:

1. $y(x) = 2 \sin x$

2. $y(x) = (1/3) \cos x$

3. $y(x) = -3 \sin x$

4. $y(x) = (-1/2) \cos x$

5. $y(x) = \sin x + 2$

6. $y(x) = \cos x - 3$

7. $y(x) = 4 \sin 2x$

8. $y(x) = -4 \cos (x/2)$

9. $y(x) = \cos (x + 30°)$

10. $y(x) = \sin (x - 90°)$

11. $y(x) = 3 \cos (2x + 180°)$

12. $y(x) = 2 \sin (x - 60°) + 1$

In problems 13 to 18, graph the indicated functions:

13. $y(x) = \tan x$

14. $y(x) = \cot x$

15. $y(x) = 1/2 \tan x$

16. $y(x) = \tan x + 2$

17. $y(x) = \cot 2x$

18. $y(x) = \tan (x - 90°)$

19. Draw the graphs of $y(x) = \sin x$ and $y(x) = \csc x$ on the same axis.

20. Draw the graphs of $y(x) = \cos x$ and $y(x) = \sec x$ on the same axis.

In problems 21 to 24, graph the indicated functions.

21. $y(x) = \csc 2x$

22. $y(x) = 2 \sec x$

23. $y(x) = \sec x + 1$

24. $y(x) = 3/2 \sec 2x + 1/2$

✓ Check Yourself

The x-axes of all of the graphs in this chapter will be drawn in degree units.

1. $y(x) = 2 \sin x$ has amplitude equal to 2 and period equal to 360°.

$$y(x) = 2 \sin x$$

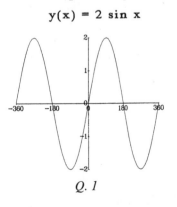

Q. 1

2. y(x) = 1/3 cos x has amplitude equal to 1/3 and period equal to 360°.

y(x) = 1/3 cos x

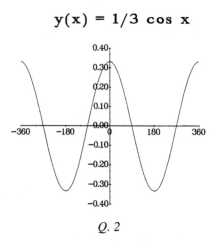

Q. 2

3. y(x) = −3 sin x has amplitude equal to 3 and period equal to 360°.

y(x) = -3 sin x

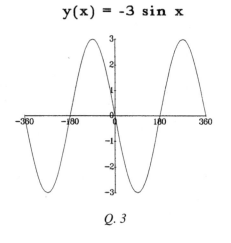

Q. 3

4. y(x) = −1/2 cos x has amplitude equal to 1/2 and period equal to 360°.

y(x) = -1/2 cos x

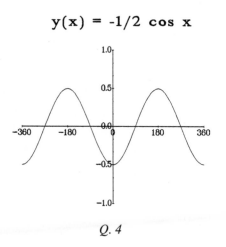

Q. 4

5. y(x) = sin x + 2 has amplitude equal to 1, period equal to 360°, and is lifted two units up.

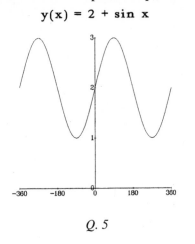

Q. 5

6. y(x) = cos x − 3 has amplitude 1, period 360°, and is pushed three units down.

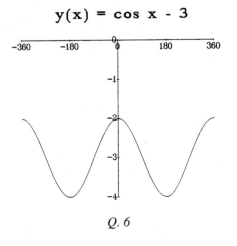

Q. 6

7. y(x) = 4 sin 2x has amplitude 4, and period 360°/2 = 180°.

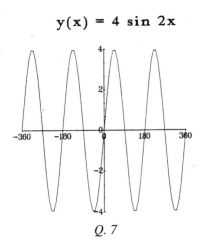

Q. 7

8. $y(x) = -4 \cos (x/2)$ has amplitude 4, and period $360°/(1/2) = 720°$.

y(x) = -4 cos(1/2)x

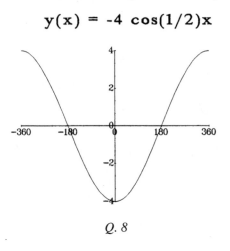

Q. 8

9. $y(x) = \cos (x + 30°)$ has amplitude 1, period 360°, and is shifted 30° to the left.

$$y(x) = \cos (x + 30°)$$

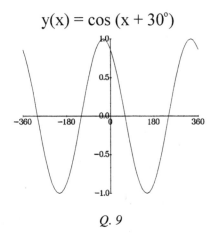

Q. 9

10. $y(x) = \sin(x - 90°)$ has amplitude 1, period 360°, and is shifted 90° to the right.

$$y(x) = \sin (x - 90°)$$

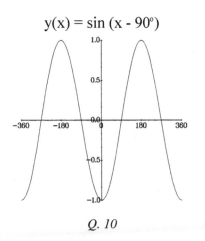

Q. 10

11. y(x) = 3 cos (2x + 180°) has amplitude 3, period of 180°, and is shifted 90° to the left.

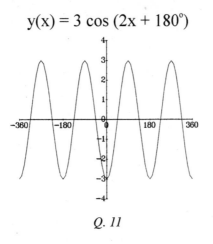

Q. 11

12. y(x) = 2 sin (x − 60°) + 1 has amplitude 2, period 360°, is shifted 60° to the right, and lifted one unit.

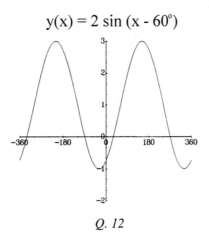

Q. 12

(Graphs of the sine and cosine functions)

13. y(x) = tan x has period 180° and asymptotes at odd multiples of 90°.

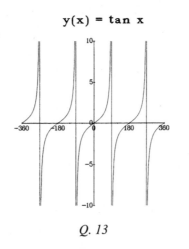

Q. 13

14. y(x) = cot x has period 180° and asymptotes at multiples of 180°.

y(x) = cot x

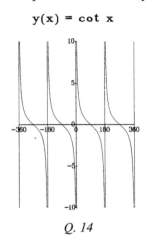

Q. 14

15. y(x) = 1/2 tan x has the same period and asymptotes as y(x) = tan x, but increases and decreases at a slower pace.

y(x) = 1/2 tan x

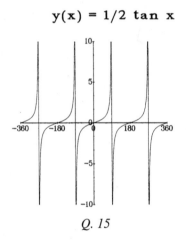

Q. 15

16. y(x) = tan x + 2 has the same period and asymptotes as y(x) = tan x, but is lifted two units.

y(x) = 2 + tan x

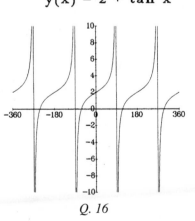

Q. 16

17. $y(x) = \cot 2x$ has a period of $180°/2 = 90°$, and asymptotes at multiples of $90°$.

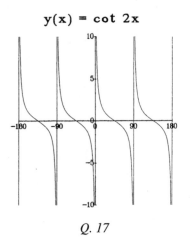

Q. 17

18. $y(x) = \tan(x - 90°)$ has a period of $180°$ and is shifted $90°$ to the right.

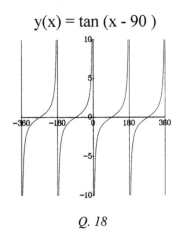

Q. 18

(Graphs of the tangent and cotangent functions)

19. In the graph below, $y(x) = \sin x$ is drawn in a heavier line than $y(x) = \csc x$. Note that every value on $y(x) = \csc x$ is the reciprocal of the corresponding value on $y(x) = \sin x$.

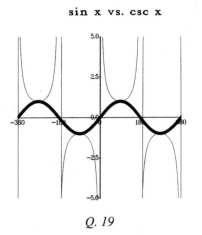

Q. 19

20. In the graph below, y(x) = cos x is drawn in a heavier line than y(x) = sec x. Note that every value on y(x) = sec x is the reciprocal of the corresponding value on y(x) = cos x.

cos x vs. sec x

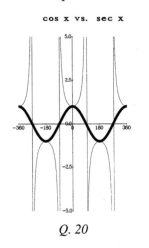

Q. 20

21. y(x) = csc 2x has period 180° and asymptotes at multiples of 90°.

y(x) = csc (2x)

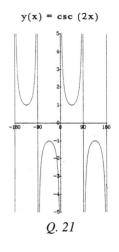

Q. 21

22. Every value of y(x) = 2 sec x is the reciprocal of the corresponding value of y(x) = 2 cos x. The period is 360°, and there are asymptotes at odd multiples of 90°.

y(x) = 2 sec x

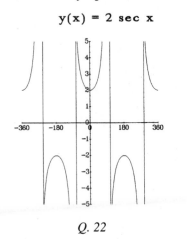

Q. 22

23.   y(x) = sec x + 1 looks like y(x) = sec x but is lifted one unit.

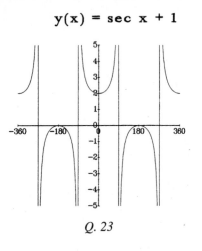

y(x) = sec x + 1

Q. 23

24.   y(x) = 3/2 sec 2x + 1/2 has a period of 180°. It looks like y(x) = 3/2 sec 2x, but is lifted horizontally 1/2 unit.

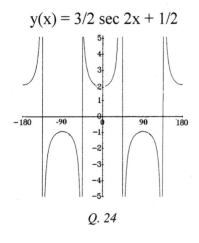

y(x) = 3/2 sec 2x + 1/2

Q. 24

(Graphs of the secant and cosecant functions)

Grade Yourself

Circle the numbers of the questions you missed, then fill in the total incorrect for each topic. If you answered more than three questions incorrectly, you need to focus on that topic. (If a topic has less than three questions and you had at least one wrong, we suggest you study that topic also. Read your textbook, a review book, or ask your teacher for help.)

Subject: Graphs of Trigonometric Functions

Topic	Question Numbers	Number Incorrect
Graphs of the sine and cosine functions	1, 2, 3, 4, 5, 6, 7, 8 , 9, 10, 11, 12	
Graphs of the tangent and cotangent functions	13, 14, 15, 16, 17, 18	
Graphs of the secant and cosecant functions	19, 20, 21, 22, 23, 24	

Inverse Trigonometric Functions

4

Brief Yourself

If f is a function such that each element of its range is associated with exactly one element of its domain, then f is called a one-to-one function. Every one-to-one function has an inverse function, denoted f^{-1}, with the property that $f(f^{-1}(x)) = f^{-1}(f(x)) = x$. The inverse function f^{-1}, essentially undoes the action of f; if you start with x and find f(x), then f^{-1} of the result will take you back to x.

The trigonometric functions are periodic, and therefore are not one-to-one in their domain. For example, $\sin(0) = \sin(\pi) = \sin(2\pi) = 0$, so the range value of 0 is associated with many values in the domain of the sine function. Thus, the trigonometric functions do not have inverse functions. However, if the domains of the trigonometric functions are restricted to certain values within which the functions *are* one-to-one, inverse functions *can* be defined. For example, for x in the interval $[-\pi/2, \pi/2]$, the function $y = \sin x$ is increasing and takes on every value from –1 to 1 exactly once. Thus, the function $y = \sin x$ with domain $[-\pi/2, \pi/2]$ is one-to-one and *does* have an inverse function.

The inverse sine function, denoted by Sin^{-1}, is defined by

$$y = \text{Sin}^{-1}x \text{ if an only if } x = \sin y, \text{ where } -\pi/2 \leq y \leq \pi/2.$$

It often helps to think of Sin^{-1} in this way: $y = \text{Sin}^{-1}x$ means that y is the number between $-\pi/2$ and $\pi/2$ whose sine is x. Thus, for example, $\text{Sin}^{-1}1/2 = \pi/6$. Since Sin^{-1} is the inverse of sin, note that $\sin(\text{Sin}^{-1}x) = x$.

Inverses for all of the other trigonometric functions can be defined by suitably restricting their domains.

A summary of definitions and properties follows.

Summary of Definitions and Properties:

$$\theta = Sin^{-1}x \Leftrightarrow x = \sin\theta \text{ and } |\theta| \le \pi/2$$

$$\theta = Cos^{-1}x \Leftrightarrow x = \cos\theta \text{ and } 0 \le \theta \le \pi$$

$$\theta = Tan^{-1}x \Leftrightarrow x = \tan\theta \text{ and } |\theta| < \pi/2$$

$$\theta = Csc^{-1}x \Leftrightarrow x = \csc\theta \text{ and } |\theta| < \pi/2$$

$$\theta = Sec^{-1}x \Leftrightarrow x = \sec\theta \text{ and } 0 < \theta < \pi$$

$$\theta = Cot^{-1}x \Leftrightarrow x = \cot\theta \text{ and } 0 < \theta < \pi$$

$$\sin(Sin^{-1}x) = x \text{ for all } x, |x| \le 1$$

$$Sin^{-1}(\sin\theta) = \theta \text{ for all } \theta, |\theta| \le \pi/2$$

$$\cos(Cos^{-1}x) = x \text{ for all } x, |x| \le 1$$

$$Cos^{-1}(\cos\theta) = \theta \text{ for all } \theta, 0 \le \theta \le \pi$$

$$\tan(Tan^{-1}x) = x \text{ for all } x$$

$$Tan^{-1}(\tan\theta) = \theta \text{ for all } \theta, |\theta| \le \pi/2$$

Test Yourself

Without using a calculator, evaluate:

1. $Sin^{-1}(\sqrt{3}/2)$

2. $Cos^{-1}(1)$

3. $Cos^{-1}(0)$

4. $Tan^{-1}(\sqrt{3})$

5. $Tan^{-1}(-1)$

6. $Sin^{-1}(-1/2)$

7. $Sin^{-1}(1)$

8. $Cos^{-1}(-1)$

9. $Cos^{-1}(-2)$

10. $Sin^{-1}(-\sqrt{2}/2)$

Use a calculator to evaluate. Give your answer in degrees to the nearest hundredth.

11. $Sin^{-1}(-.76)$

12. $Cos^{-1}(.8890)$

13. $Tan^{-1}(2.2)$

14. $Sin^{-1}(.3)$

15. $Cos^{-1}(-.011)$

16. $Tan^{-1}(-.855)$

Simplify the following expressions:

17. $\sin(Sin^{-1}(.789))$

18. $\tan(Tan^{-1}(-22.1))$

19. $Sin^{-1}(\sin(2\pi/3))$

20. $Cos^{-1}(\cos(4\pi/3))$

21. $\cos(Cos^{-1}(-.3751))$

22. $\cos(Cos^{-1}(0))$

23. $\tan(Tan^{-1}(.001))$

24. $\sin(Sin^{-1}(-2.6))$

25. $\tan(Tan^{-1}(761))$

26. $Sin^{-1}(\sin(-\pi/3))$

27. $Cos^{-1}(\cos \pi)$

28. $Sin^{-1}(\sin \pi)$

29. $Cos^{-1}(\cos(-\pi/3))$

30. $Cos^{-1}(\cos 2\pi)$

31. $Tan^{-1}(\tan 2\pi)$

32. $Tan^{-1}(\tan(5\pi/4))$

33. $Sin^{-1}(\sin(7\pi/6))$

34. $Cos^{-1}(\cos(3\pi/7))$

35. $Cos^{-1}(\cos(13\pi/10))$

36. $Tan^{-1}(\tan(-5\pi/6))$

Evaluate the following expressions:

37. $\sin(Cos^{-1}(3/5))$

38. $\cos(Tan^{-1}(-3/4))$

39. $\sin(Cos^{-1}(12/13))$

40. $\cos(Sin^{-1}(12/13))$

41. $\sin(Tan^{-1}(-5/12))$

42. $\tan(Sin^{-1}(-3/5))$

43. $\tan(Cos^{-1}(-3/5))$

44. $\sin(Tan^{-1}(1/\sqrt{3}))$

45. $\cos(Tan^{-1}(-1/\sqrt{3}))$

46. $\cos(Sin^{-1}(1))$

47. $\sin(Cos^{-1}(-1))$

48. $\sin(Tan^{-1}(1))$

49. $\sin(Tan^{-1}(-1))$

50. $\cos(Tan^{-1}(1/\sqrt{2}))$

51. $\cos(Tan^{-1}(-1/\sqrt{2}))$

52. $\tan(Sin^{-1}2)$

53. $\sin(Tan^{-1}2)$

54. $\tan(Cos^{-1}(-3))$

55. $\cos(Tan^{-1}(-3))$

Without using a calculator, evaluate:

56. $Sec^{-1}2$

57. $Csc^{-1}(2/\sqrt{3})$

58. $Cot^{-1}(-1)$

59. $Csc^{-1}(\sqrt{2})$

60. $Sec^{-1}(-\sqrt{2})$

61. $Cot^{-1}(\sqrt{3})$

62. $Cot^{-1}(1/\sqrt{3})$

63. $Sec^{-1}(-2)$

64. $Csc^{-1}(-2)$

Use a calculator to evaluate. Give your answer in degrees to the nearest hundredth.

65. $Sec^{-1}(2.2)$

66. $Csc^{-1}(-2.6)$

67. $Cot^{-1}(4)$

68. $Cot^{-1}(-4)$

69. $Sec^{-1}(-2.2)$

70. $Csc^{-1}(2.6)$

Evaluate the following expressions:

71. $\tan(Sec^{-1}(-3))$

72. $Sec^{-1}(\sec(4\pi/3))$

73. $Cot^{-1}(\cot(-5\pi/4))$

74. $Csc^{-1}(\csc(5\pi/6))$

75. $\sin(Cot^{-1}(3/4))$

76. $\cos(Cot^{-1}(-3/4))$

77. $\sin(Cot^{-1}(4/3))$

78. $\cos(Cot^{-1}(-4/3))$

79. $\tan(Sec^{-1}(-2/\sqrt{3}))$

80. $\tan(Sec^{-1}(2/\sqrt{3}))$

81. $\csc(Sec^{-1}(2/\sqrt{3}))$

✓ Check Yourself

1. For $\theta = \pi/3$, $\sin \theta = \sqrt{3}/2$ and $|\theta| \leq \pi/2$ (**Evaluation of inverse functions**)

2. For $\theta = 0$, $\cos \theta = 1$, and $0 \leq \theta \leq \pi$ (**Evaluation of inverse functions**)

3. For $\theta = \pi/2$, $\cos \theta = 0$ and $0 \leq \theta \leq \pi$ (**Evaluation of inverse functions**)

4. For $\theta = \pi/3$, $\tan \theta = \sqrt{3}$ and $|\theta| \leq \pi/2$ (**Evaluation of inverse functions**)

5. For $\theta = -\pi/4$, $\tan \theta = -1$ and $|\theta| \leq \pi/2$ (**Evaluation of inverse functions**)

6. For $\theta = -\pi/6$, $\sin \theta = -1/2$ and $|\theta| \leq \pi/2$ (**Evaluation of inverse functions**)

7. For $\theta = \pi/2$, $\sin \theta = 1$ and $|\theta| \leq \pi/2$ (**Evaluation of inverse functions**)

8. For $\theta = \pi$, $\cos \theta = -1$ and $0 \leq \theta \leq \pi$ (**Evaluation of inverse functions**)

9. $Cos^{-1}(-2)$ is undefined   (**Evaluation of inverse functions**)

10. For $\theta = -\pi/4$, $\sin \theta = -\sqrt{2}/2$ and $|\theta| \leq \pi/2$ (**Evaluation of inverse functions**)

11. $Sin^{-1}(-.76) = -49.46°$ (**Evaluation of inverse functions**)

12. $Cos^{-1}(.8890) = 27.25°$ (**Evaluation of inverse functions**)

13. $Tan^{-1}(2.2) = 65.56°$ (**Evaluation of inverse functions**)

14. $Sin^{-1}(.3) \cong 17.46°$ **(Evaluation of inverse functions)**

15. $Cos^{-1}(-.011) = 90.63°$ **(Evaluation of inverse functions)**

16. $Tan^{-1}(-.855) = -40.53°$ **(Evaluation of inverse functions)**

17. $\sin(Sin^{-1}(.789)) = .789$ since $\sin(Sin^{-1}(x)) = x$ for $-1 \le x \le 1$.   **(Simplification of compound expressions)**

18. $\tan(Tan^{-1}(-22.1)) = -22.1$ since $\tan(Tan^{-1}(x)) = x$ for all x.   **(Simplification of compound expressions)**

19. $Sin^{-1}(\sin(2\pi/3))$ does not equal $2\pi/3$ since $2\pi/3 > \pi/2$.

    However, $\sin(2\pi/3) = \sin(\pi/3)$ and $-\pi/2 < \pi/3 \le \pi/2$.

    Therefore $Sin^{-1}(\sin(2\pi/3)) = Sin^{-1}(\sin(\pi/3)) = \pi/3$.   **(Simplification of compound expressions)**

20. $Cos^{-1}(\cos(4\pi/3))$ does not equal $4\pi/3$ since $4\pi/3 > \pi$.

    But $\cos(4\pi/3) = \cos(2\pi/3)$ and $0 < 2\pi/3 < \pi$.

    Thus $Cos^{-1}(\cos(4\pi/3)) = Cos^{-1}(\cos(2\pi/3)) = 2\pi/3$.   **(Simplification of compound expressions)**

21. $\cos(Cos^{-1}(-.3751)) = -.3751$ since $-1 < -.3751 < 1$.   **(Simplification of compound expressions)**

22. $\cos(Cos^{-1}(0)) = 0$ since $-1 < 0 < 1$.   **(Simplification of compound expressions)**

23. $\tan(Tan^{-1}(.001)) = .001$ (Note that $\tan(Tan^{-1}(x)) = x$ for all values of x.)   **(Simplification of compound expressions)**

24. The value -2.6 is not in the domain of the inverse sine function; hence, there is no angle with sine equal to -2.6. The expression is meaningless.   **(Simplification of compound expressions)**

25. $\tan(Tan^{-1}(761)) = 761$ (Note that $\tan(Tan^{-1}(x)) = x$ for all values of x.)   **(Simplification of compound expressions)**

26. $Sin^{-1}(\sin(-\pi/3)) = -\pi/3$ since $-\pi/2 \le -\pi/3 \le \pi/2$.   **(Simplification of compound expressions)**

27. $Cos^{-1}(\cos(\pi)) = \pi$ since $0 \le \pi \le \pi$.   **(Simplification of compound expressions)**

28. π is not in the interval $[-\pi/2, \pi/2]$ but $\sin \pi = \sin 0 = 0$ and $-\pi/2 \le 0 \le \pi/2$.   **(Simplification of compound expressions)**

    Therefore $Sin^{-1}(\sin \pi) = Sin^{-1}(0) = 0$.   **(Simplification of compound expressions)**

29. The interval $[0,\pi]$ does not contain $-\pi/3$ but $\cos(-\pi/3) = \cos(\pi/3)$ and $0 \le \pi/3 \le \pi$.

    Therefore $Cos^{-1}(\cos(-\pi/3)) = Cos^{-1}(\cos(\pi/3)) = \pi/3$.   **(Simplification of compound expressions)**

30. The interval $[0,\pi]$ does not contain 2π but $\cos(2\pi) = \cos(0)$ and $0 \le 0 \le \pi$.

    Therefore $Cos^{-1}(\cos(2\pi)) = Cos^{-1}(\cos(0)) = 0$.   **(Simplification of compound expressions)**

31. The interval $[-\pi/2, \pi/2]$ does not contain 2π but $\tan(2\pi) = \tan(0)$ and $-\pi/2 \le 0 \le \pi/2$.

    Therefore $Tan^{-1}(\tan(2\pi)) = Tan^{-1}(\tan(0)) = 0$.   **(Simplification of compound expressions)**

32. The interval $[-\pi/2, \pi/2]$ does not contain $5\pi/4$ but $\tan(5\pi/4) = \tan(\pi/4)$ and $-\pi/2 \le \pi/4 \le \pi/2$.

Therefore $Tan^{-1}(\tan(5\pi/4)) = Tan^{-1}(\tan(\pi/4)) = \pi/4$.   **(Simplification of compound expressions)**

33. The interval $[-\pi/2, \pi/2]$ does not contain $7\pi/6$ but $\sin(7\pi/6) = \sin(-\pi/6)$ and $-\pi/2 \le -\pi/6 \le \pi/2$.

Therefore $Sin^{-1}(\sin(7\pi/6)) = Sin^{-1}(\sin(-\pi/6)) = -\pi/6$.   **(Simplification of compound expressions)**

34. Since $0 \le 3\pi/7 \le \pi$, $Cos^{-1}(\cos(3\pi/7)) = 3\pi/7$.   **(Simplification of compound expressions)**

35. The interval $[0, \pi]$ does not contain $13\pi/10$ but $\cos(13\pi/10) = \cos(7\pi/10)$ and $0 \le 7\pi/10 \le \pi$.

Therefore $Cos^{-1}(\cos(13\pi/10)) = Cos^{-1}(\cos(7\pi/10)) = 7\pi/10$.   **(Simplification of compound expressions)**

36. The interval $[-\pi/2, \pi/2]$ does not contain $-5\pi/6$ but $\tan(-5\pi/6) = \tan(\pi/6)$ and $-\pi/2 \le \pi/6 \le \pi/2$.

Therefore $Tan^{-1}(\tan(-5\pi/6)) = Tan^{-1}(\tan(\pi/6)) = \pi/6$.   **(Simplification of compound expressions)**

37. Since $0 < 3/5 < 1$, the angle $Cos^{-1}(3/5)$ is in quadrant I. A first quadrant angle with cosine equal to $3/5$ is shown here.

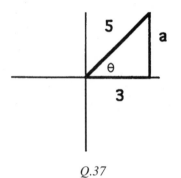

Q.37

Then

$\sin\theta = a/5$

and $a = \sqrt{5^2 - 3^2} = 4$.

Thus, $\sin(Cos^{-1}(3/5)) = 4/5$.   **(Simplification of compound expressions)**

38. Since $-3/4 < 0$, the angle $Tan^{-1}(3/4)$ is in quadrant IV, between $-\pi/2$ and 0. A fourth quadrant angle θ with tangent equal to $-3/4$ is shown here.

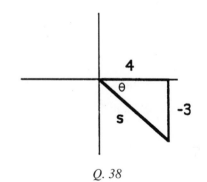

Q. 38

Then

$\cos\theta = 4/s$

and $s = \sqrt{3^2 + 4^2} = 5$.

Thus, $\cos(Tan^{-1}(-3/4)) = 4/5$.   **(Simplification of compound expressions)**

39. Since $0 < 12/13 < 1$, the angle $Cos^{-1}(12/13)$ is in quadrant I. A first quadrant angle θ with cosine equal to 12/13 is shown here.

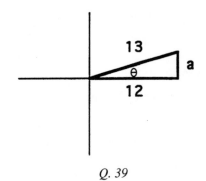

Q. 39

Then

$\sin \theta = a/13$

and $a = \sqrt{13^2 - 12^2} = 5.$

Thus, $\sin(Cos^{-1}(12/13)) = 5/13.$ **(Simplification of compound expressions)**

40. Since $0 < 12/13 < 1$, the angle $Sin^{-1}(12/13)$ is in quadrant I. A first quadrant angle θ with sine equal to 12/13 is shown here.

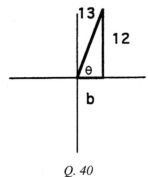

Q. 40

Then

$\cos \theta = b/13$

and $b = \sqrt{13^2 - 12^2} = 5.$

Thus, $\cos(Sin^{-1}(12/13)) = 5/12.$ **(Simplification of compound expressions)**

41. Since $-5/12 < 0$, the angle $Tan^{-1}(-5/12)$ is in quadrant IV, between $-\pi/2$ and 0. A fourth quadrant angle θ with tangent equal to $-5/12$ is shown here.

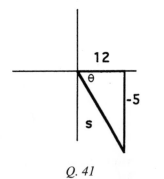

Q. 41

Then

$\sin \theta = 5/s$

and $s = \sqrt{5^2 + 12^2} = 13$.

Thus, $Sin(Tan^{-1}(-5/12)) = -5/13$.   **(Simplification of compound expressions)**

42. Since $-3/5 < 0$, the angle $Sin^{-1}(-3/5)$ is in quadrant IV, between $-\pi/2$ and 0. A fourth quadrant angle θ with sine equal to $-3/5$ is shown here.

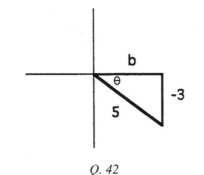

Q. 42

Then

$\tan \theta = -3/b$

and $b = \sqrt{5^2 - 3^2} = 4$.

Thus, $\tan(Sin^{-1}(-3/5)) = -3/4$.   **(Simplification of compound expressions)**

43. Since $-3/5 < 0$, the angle $Cos^{-1}(-3/5)$ is in quadrant II. A second quadrant angle θ with cosine equal to $-3/5$ is shown here.

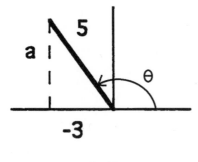

Q. 43

Then

$\tan \theta = a/(-3)$

and $a = \sqrt{5^2 - (-3)^2} = 4$.

Thus, $\tan(Cos^{-1}(-3/5)) = -4/3$.   **(Simplification of compound expressions)**

44. Since $1/\sqrt{3} > 0$, the angle $Tan\text{-}1(1/\sqrt{3})$ is in quadrant I. A first quadrant angle θ with tangent equal to $1/\sqrt{3}$ is shown here.

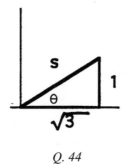

Q. 44

Then

$\sin \theta = 1/s$

and $s = \sqrt{1^2 + 3} = 2.$

Thus, $\sin(Tan^{-1}(1/\sqrt{3})) = 1/2.$ **(Simplification of compound expressions)**

45. Since $-1/\sqrt{3} > 0$, the angle $Tan^{-1}(-1/\sqrt{3})$ is in quadrant IV, between $-\pi/2$ and 0. A fourth quadrant angle θ with tangent equal to $-1/\sqrt{3}$ is shown here.

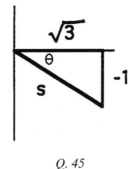

Q. 45

Then

$\cos \theta = \sqrt{3}/s$

and $s = \sqrt{\sqrt{3}^2 + 1} = 2.$

Thus, $\cos(Tan^{-1}(-1/\sqrt{3})) = \sqrt{3}/2.$ **(Simplification of compound expressions)**

46. The range of the inverse sine function is the interval $[-\pi/2, \pi/2]$ and since $1 > 0$, the angle $Sin^{-1}(1)$ is $\pi/2$. In fact $\sin(\pi/2) = 1$ and $\cos(\pi/2) = 0$.

Hence, $\cos(Sin^{-1}1) = \cos(\pi/2) = 0.$ **(Simplification of compound expressions)**

47. The range of the inverse cosine function is the interval $[0, \pi]$ and since $-1 < 0$, the angle $Cos^{-1}(\text{-}1)$ is π. In fact $\cos(\pi) = -1$ and $\sin(\pi) = 0$.

Hence, $\sin(Cos^{-1}(-1)) = \sin \pi = 0.$ **(Simplification of compound expressions)**

48. The range of the inverse tangent function is the interval $[-\pi/2, \pi/2]$ and since $1 > 0$, the angle $Tan^{-1}(1)$ is in the first quadrant. In fact $\tan(\pi/4) = 1$ and $\sin(\pi/4) = \sqrt{2}/2$.

     Then $\sin(Tan^{-1}1) = \sin \pi/4 = \sqrt{2}/2.$ **(Simplification of compound expressions)**

49. Since $-1 < 0$, $Tan^{-1}(-1)$ lies in the interval $[-\pi/2, 0]$. In fact $\tan(-\pi/4) = -1$ and $\sin(-\pi/4) = -\sqrt{2}/2$.

     Hence, $\sin(Tan^{-1}(-1)) = \sin(-\pi/4) = -\sqrt{2}/2.$ **(Simplification of compound expressions)**

50. Since $1/\sqrt{2} > 0$, the angle $Tan^{-1}(1/\sqrt{2})$ is in quadrant I, between 0 and $\pi/2$. A first quadrant angle θ with tangent equal to $1/\sqrt{2}$ is shown here.

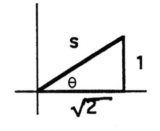

Q. 50

Then

$\cos \theta = \sqrt{2} / s$

and $s = \sqrt{1^2 + \sqrt{2}^2} = \sqrt{3}$.

Thus, $\cos(Tan^{-1}(1/\sqrt{2})) = \sqrt{2} / \sqrt{3}.$ **(Simplification of compound expressions)**

51. Since $-1/\sqrt{2} < 0$, the angle $Tan^{-1}(-1/\sqrt{2})$ is in quadrant IV, between $-\pi/2$ and 0. A fourth quadrant angle θ with tangent equal to $-1/\sqrt{2}$ is shown here.

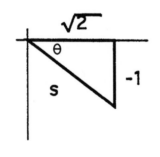

Q. 51

Then

$\cos \theta = \sqrt{2} / s$

and $s = \sqrt{1^2 + \sqrt{2}^2} = \sqrt{3}$.

Thus, $\cos(Tan^{-1}(-1/\sqrt{2})) = \sqrt{2} / \sqrt{3}.$ **(Simplification of compound expressions)**

52. The range of the sine function is -1 to 1; there is no angle whose sine equals 2. This problem has no solutions. **(Simplification of compound expressions)**

53. Since $2 > 0$, the angle $Tan^{-1}(2)$ is in quadrant I, between 0 and $\pi/2$. A first quadrant angle θ with tangent equal to 2 is shown here.

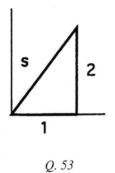

Q. 53

Then

$\text{Sin } \theta = 2 \,/\, s$

and $s = \sqrt{1^2 + 2^2} = \sqrt{5}$.

Thus, $\sin(Tan^{-1}2 = 2\,/\!\sqrt{5})$.   **(Simplification of compound expressions)**

54. The range of the cosine function is -1 to 1; there is no angle whose cosine equals -3. This problem has no solutions.   **(Simplification of compound expressions)**

55. Since $-3 < 0$, the angle $Tan^{-1}(-3)$ is in quadrant IV, between $-\pi/2$ and 0. A fourth quadrant angle θ with tangent equal to -3 is shown here.

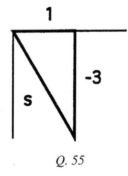

Q. 55

Then

$\cos \theta = 1 \,/\, s$

and $s = \sqrt{1^2 + (-3)^2} = \sqrt{10}$.

Thus, $\cos(Tan^{-1}(-3)) = 1 \,/\, \sqrt{10}$.   **(Simplification of compound expressions)**

56. $x = Sec^{-1}2$ if x is between 0 and π and sec x = 2. *Sin*ce sec x = 2 if and only if cos x = 1/2, it follows that $x = \pi/3$.   **(Evaluation of inverse functions)**

57. $x = Csc^{-1}(2/\sqrt{3})$ if x is between $-\pi/2$ and $\pi/2$ and csc x = $2/\sqrt{3}$. *Sin*ce csc x = $2/\sqrt{3}$ if and only if sin x = $\sqrt{3}/2$ it follows that x = $\pi/3$.   **(Evaluation of inverse functions)**

58. $x = Cot^{-1}(-1)$ if x is between 0 and π and cot x = -1. Note that cot x = -1 if and only if tan x = -1, hence x = $3\pi/4$.   **(Evaluation of inverse functions)**

59. $x = Csc^{-1}(\sqrt{2})$ if x is between $-\pi/2$ and $\pi/2$ and csc x = $\sqrt{2}$. *Since* csc x = $\sqrt{2}$ if and only if sin x = $1/\sqrt{2}$, it follows that x = $\pi/4$. **(Evaluation of inverse functions)**

60. $x = Sec^{-1}(-\sqrt{2})$ if x is between 0 and π and sec x = $-\sqrt{2}$. In this case cos x = $-1/\sqrt{2}$, hence x = $3\pi/4$. **(Evaluation of inverse functions)**

61. $x = Cot^{-1}(\sqrt{3})$ if x is between 0 and π and cot x = $\sqrt{3}$. In this case tan x = $1/\sqrt{3}$, hence x = $\pi/6$. **(Evaluation of inverse functions)**

62. $x = Cot^{-1}(1/\sqrt{3})$ if x is between 0 and π and cot x = $1/\sqrt{3}$. In this case tan x = $\sqrt{3}$, hence x = $\pi/3$. **(Evaluation of inverse functions)**

63. $x = Sec^{-1}(-2)$ if x is between 0 and π and sec x = -2. Since sec x = -2 if and only if cos x = $-1/2$, it follows that x = $2\pi/3$. **(Evaluation of inverse functions)**

64. $x = Csc^{-1}(-2)$ if x is between $-\pi/2$ and $\pi/2$ and csc x = -2. Since csc x = -2 if and only if sin x = $-1/2$, it follows that x = $-\pi/6$. **(Evaluation of inverse functions)**

65. $Sec^{-1}(2.2) = Cos^{-1}(1/2.2) = 62.96°$ **(Evaluation of inverse functions)**

66. $Csc^{-1}(-2.6) = Sin^{-1}(-1/2.6) = -22.62°$ **(Evaluation of inverse functions)**

67. $Cot^{-1}(4) = Tan^{-1}(1/4) = 14.04°$ **(Evaluation of inverse functions)**

68. $Cot^{-1}(-4) = Tan^{-1}(-1/4) = -Tan^{-1}(1/4) = 165.96°$ **(Evaluation of inverse functions)**

69. $Sec^{-1}(-2.2) = Cos^{-1}(-1/2.2) = 117.04°$ **(Evaluation of inverse functions)**

70. $Csc^{-1}(2.6) = Sin^{-1}(1/2.6) = 22.62°$ **(Evaluation of inverse functions)**

71. $x = Sec^{-1}(-3)$ if x is between $\pi/2$ and π and sec x = -3. Such an angle is shown here.

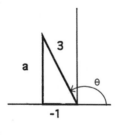

Q. 71

Then

$\tan(Sec^{-1}(-3)) = a / (-1) = -a$

where $a = \sqrt{3^2 - (-1)^2} = \sqrt{8}$.

Thus, $\tan(Sec^{-1}(-3)) = -\sqrt{8}.$ **(Simplification of compound expressions)**

72. Since x=4π/3 is not in the interval [0,π], the inverse secant of sec x is not equal to x. But sec(4π/3) = sec(2π/3) and 2π/3 lies in the interval [0,π].

    Then

    $Sec^{-1}(\sec(4\pi/3)) = Sec^{-1}(\sec(2\pi/3)) = 2\pi/3.$ **(Simplification of compound expressions)**

73. Since x= – 5π/4 is not in the interval [0,π], the inverse cotangent of cot x is not equal to x. But cot(– 5π/4) = cot(3π/4) and 3π/4 lies in the interval [0,π].

    Then

    $Cot^{-1}(\cot(-5\pi/4)) = Cot^{-1}(\cot(3\pi/4)) = 3\pi/4.$ **(Simplification of compound expressions)**

74. Since x=5π/6 is not in the interval [– π/2,π/2], the inverse cosecant of csc x is not equal to x. But csc(5π/6) = csc(π/6) and π/6 lies in the interval [– π/2,π/2].

    Then

    $Csc^{-1}(\csc(5\pi/6)) = Csc^{-1}(\csc(\pi/6)) = \pi/6.$ **(Simplification of compound expressions)**

75. The inverse cotangent of 3/4 is an angle x, in [0,π/2] with cot x = 3/4.

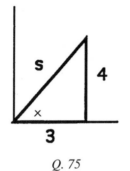

Q. 75

    Then

    $\sin x = 4/s$

    where $s = \sqrt{3^2 + 4^2} = 5.$

    Thus $\sin(Cot^{-1}(3/4)) = 4/5.$ **(Simplification of compound expressions)**

76. The inverse cotangent of – 3/4 is an angle x, in [π/2,π] with cot x = – 3/4.

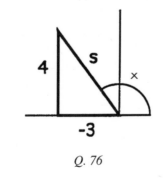

Q. 76

    Then

    $\sin x = 4/s$

where $s = \sqrt{3^2 + 4^2} = 5$.

Thus, $\cos(Cot^{-1}(-3/4)) = -3/5$. **(Simplification of compound expressions)**

77. The inverse cotangent of 4/3 is an angle x, in [0,π/2] with cot x = 4/3.

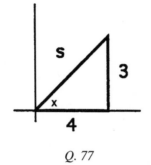

Q. 77

Then

$\sin x = 3 / s$

where $s = \sqrt{3^2 + 4^2} = 5$.

Thus, $\sin(Cot^{-1}(4/3)) = 3/5$. **(Simplification of compound expressions)**

78. The inverse cotangent of – 4/3 is an angle x, in [π/2,π] with cot x = – 4/3.

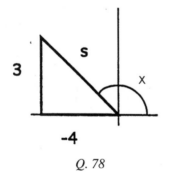

Q. 78

Then

$\cos x = -4 / s$

where $s = \sqrt{3^2 + 4^2} = 5$.

Thus, $\cos(Cot^{-1}(-4/3)) = -4/5$. **(Simplification of compound expressions)**

79. The inverse secant of $-2/\sqrt{3}$ is an angle x, in $[\pi/2,\pi]$ with sec x $= -2/\sqrt{3}$.

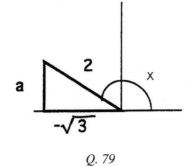

Q. 79

Then

$\tan x = a/(-\sqrt{3})$

where $a = \sqrt{2^2-(-\sqrt{3})^2} = 1$.

Thus, $\tan(Sec^{-1}(-2/\sqrt{3})) = 1/(-\sqrt{3})$. **(Simplification of compound expressions)**

80. The inverse secant of $2/\sqrt{3}$ is an angle x, in $[0,\pi/2]$ with sec x $= 2/\sqrt{3}$.

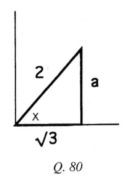

Q. 80

Then

$\tan x = a/(\sqrt{3})$

where $a = \sqrt{2^2-(\sqrt{3})^2} = 1$.

Thus, $\tan(Sec^{-1}(2/\sqrt{3})) = 1/(\sqrt{3})$. **(Simplification of compound expressions)**

81. By the result of the previous problem, $\csc(Sec^{-1}(2/\sqrt{3})) = 2/a = 2$. **(Simplification of compound expressions)**

Grade Yourself

Circle the numbers of the questions you missed, then fill in the total incorrect for each topic. If you answered more than three questions incorrectly, you need to focus on that topic. (If a topic has less than three questions and you had at least one wrong, we suggest you study that topic also. Read your textbook, a review book, or ask your teacher for help.)

Subject: Inverse Trigonometric Functions

Topic	Question Numbers	Number Incorrect
Evaluation of inverse functions	1, 2, 3, 4, 5, 6, 7, 8, 9, 10, 11, 12, 13, 14, 15, 16, 56, 57, 58, 59, 60, 61, 62, 63, 64, 65, 66, 67, 68, 69, 70	
Simplification of compound expressions	17, 18, 19, 20, 21, 22, 23, 24, 25, 26, 27, 28, 29, 30, 31, 32, 33, 34, 35, 36, 37, 38, 39, 40, 41, 42, 43, 44, 45, 46, 47, 48, 49, 50, 51, 52, 53, 54, 55, 71, 72, 73, 74, 75, 76, 77, 78, 79, 80, 81	

Solving Triangles

5

Brief Yourself

The sum of the angles in any plane triangle equals 180°. The sum of the two acute angles in any right triangle equals 90°. In a right triangle, the side opposite the right angle is called the hypotenuse and the other two sides are called legs. The Pythagorean theorem asserts that the sum of the squares of the legs equals the square of the hypotenuse. All the sides and all the angles of any right triangle can be found given two sides or one side and one acute angle. This is called solving the triangle.

In a general triangle (i.e., a triangle which is not a right triangle), the sides and angles are related by the law of sines and the law of cosines (see problems 4 and 5). To solve a general triangle given one side and two angles, use A+B+C = 180° to find the remaining angle and use the law of sines to find the sides. Solving a general triangle given two sides and an opposite angle also makes use of the law of sines, but is more complicated because the problem may have no solutions, two solutions, or a unique solution depending on the values of the sides and angle given. This is referred to as the ambiguous case of the law of sines. The law of cosines is used to solve general triangles in the cases when the given information consists of two sides and the included angle or when three sides are given. Summarizing these cases:

1 side, 2 angles	Law of Sines
2 sides, an opposite angle	Law of Sines (ambiguous case)
2 sides, included angle	Law of Cosines
3 sides	Law of Cosines

In order to determine the number of solutions in the ambiguous case, assume you are given angle A and sides *a* and *b*. Then,

If the measure of A ≥ 90°, there is one solution if a > b, and no solution if a ≤ b.

If the measure of A < 90°, and a ≥ b, there is one solution.

If the measure of A < 90°, and a < b, there is one solution if sin β = 1,
    no solutions is sin β > 1, and two solutions if sin β < 1.

Test Yourself

The three angles of a triangle are labeled A, B, C, and the corresponding opposite sides are labeled a,b,c, respectively.

1. What is a relationship involving A,B,C?

2. What is a right triangle and how are the angles related in a right triangle?

3. If C is a right angle, then how are a,b,c related?

4. State the law of cosines. Is this result valid in general triangles or only in right triangles?

5. State the law of sines. Is this result valid in general triangles or only in right triangles?

The angle C in triangle ABC is a right angle. In the following problems, use the given information to find any missing sides and angles:

6. a = 60,   B = 60°

7. a = 12,   b = 5

8. a = 52,   A = 45°

9. c = 21.2,   b = 11.2

10. c = 80,   A = 52°

11. c = 89,   a = 41

12. b = 60,   B = 30°

13. c = 132,   A = 16°

14. b = 31.3,   A = 67.8°

15. a = 11.5,   b = 22.7

16. c = 112.5,   B = 32.17°

17. a = 15.5,   B = 13.1°

18. b = 10,   B = 81°

19. b = 10,   A = 81°

20. c = 10,   A = 81°

21. A lamp post is 20m high. At a distance d from the base of the lamp post, the angle of elevation to the top of the post is 22°. Find the distance d.

22. At a distance of 46m from the base of a building, the angle of elevation to the top is 77°. How tall is the building?

23. A boulder at point B on the east side of a wide canyon is directly opposite a cactus at point C on the west side of the canyon. From point C, pacing off 100m along a line that is at right angles to the line from B to C, you find yourself at point A. If the line of sight from A to B makes an angle of 64° with the line from C to A, how wide is the canyon?

24. At a distance d from the base of a mountain, you measure the angle of elevation to the top to be 12.7°. After walking 400m closer to the mountain, you measure the angle of elevation to be 13.1°. How high is the mountain and how far away is it from the point where you made your first measurement?

25. A straight highway rises from an elevation of 4100 feet above sea level to an elevation of 5400 feet above sea level in a distance of 11 miles. What is the angle at which the highway is rising?

The triangle ABC is not necessarily a right triangle. Determine the missing parts of the triangle from the given information:

26. A = 70°,   B = 55°,   a = 12

27. A = 50°,   C = 33.5°,   b = 76

28. B = 107°,   C = 30.5°,   c = 126

29. A = 35°,   B = 25°,   c = 67.6

30. B = 63°,   C = 74.2°,   a = 1.1

31. A = 33°,   C = 62°,   b = 41.4

32. A = 40°,   a = 20,   b = 15

33. A = 23°,   a = 54.3,   b = 22.1

34. A = 125°,   a = 40,   b = 35

35. A = 63°,   a = 10,   c = 8.9

36. A = 75°,   a = 51,   b = 71

37. A = 136°,   a = 57.5,   c = 49.8

38. A = 37°,   a = 12,   b = 16.1

39. A = 29°,   a = 21.33,   b = 44

40. A = 58°,   a = 22,   c = 24.1

41. A = 12°,   a = 7,   b = 28

42. A = 18°,   a = 9.3,   b = 41

43. A = 49°,   a = 95,   c = 125

44. A = 112°,   a = 42.1,   c = 37

45. A = 162°,   a = 6.1,   b = 4

46. Two observers on the ground, separated by 850 feet, observe a hot air balloon in the air above them and on a line between them. The angles of elevation of the balloon are 57.3° and 44° respectively for the two observers. How high is the balloon?

47. A leaning wall is inclined 6° from the vertical. On the side of the wall which forms an acute angle with the ground and at a distance of 40 feet from the wall, the angle of elevation to the top of the wall is 22°. How high is the wall?

48. A pier extends out from a straight dock at an angle of 85°. On the side of the dock which forms an obtuse angle with the pier, the line of sight to the tip of the pier from a point 100 feet away makes an angle of 37°. How long is the pier?

49. Two buildings of equal height are 800 feet apart. An observer at point C on the street between the two buildings notes that the lines of sight to points A and B on the tops of the two buildings make angles of 27° and 41° respectively. If A, B, and C are assumed to lie in a plane perpendicular to the ground, how high are the buildings?

The triangle ABC is not necessarily a right triangle. Determine the missing parts of the triangle from the given information:

50. b = 29,   c = 17,   A = 103°

51. a = 5,   b = 7,   c = 10

52. a = 100,   b = 30,   C = 40°

53. a = 7,   b = 24,   c = 26

54. a = 70,   b = 82,   C = 42°

55. b = 91,   c = 67,   A = 49°

56. a = 17,   c = 12,   B = 66°. Find b.

57. a = 11.7,   b = 6.6,   c = 14.2. Find the smallest angle.

58. a = 36.5,   b = 40.2,   c = 26.1. Find the largest angle.

59. b = 11.2,   c = 48.2   A = 162°. Find a.

In the following problems, first determine the method to use (law of sines, law of cosines, etc.) and then find the missing parts of the triangle.

60. b = 37,   c = 11.9,   B = 90°

61. A = 18°,   C = 47°,   c = 33.1

62. A = 58°,   b = 7,   c = 14

63. a = 17,   b = 30.3,   C = 71°

64. a = 20,   b = 26,   A = 39°

✔ Check Yourself

1. $A+B+C = 180°$

2. Any triangle containing one right angle is called a right triangle. If C is a right angle, then $A+B = 90°$. A and B are said to be complementary angles. **(Properties of triangles)**

3. If C is a right angle then c is called the hypotenuse of the right triangle and a,b are its legs. The sides are related by the Pythagorean theorem which states that $c^2 = a^2 + b^2$. **(Properties of triangles)**

4. The sides and angles of any triangle satisfy the law of cosines:

   $$c^2 = a^2 + b^2 - 2ab \cos C$$
   $$b^2 = c^2 + a^2 - 2ac \cos B$$
   $$a^2 = b^2 + c^2 - 2bc \cos A \quad \textbf{(Properties of triangles)}$$

5. The sides and angles of any triangle satisfy the law of sines:

   $$\frac{\sin A}{a} = \frac{\sin B}{b} = \frac{\sin C}{c} \quad \textbf{(Properties of triangles)}$$

6. $B = 60°$ and $C = 90°$ so $A = 180°–60°–90° = 30°$.

   Alternatively, A and B are complementary so $A = 90° – 60° = 30°$.

   Since $a = 60$, $c = 60 \div \sin 30° = 60 \csc 30° = 120$

   and $b = 60 \tan 60° = 60 \sqrt{3}$.   **(Side plus angle)**

7. $a = 12$ and $b = 5$. Therefore

   $$c = \sqrt{a^2 + b^2} = \sqrt{169} = 13$$

   $$\sin A = \frac{a}{c} = \frac{12}{13} = .9231$$

   $$\sin B = \frac{b}{c} = \frac{5}{13} = .3846.$$

   Using the calculator, $A = \text{Sin}^{-1}(12/13) = 67.3801°$

   and $\qquad\qquad\qquad B = 90° – A = 22.6199°$.

   Alternatively, $\quad B = \text{InvSin}(5/13) = 22.6199°$. **(Two sides)**

8. $a = 52$ and $A = 45°$. Then $B = 90° – A = 45°$ and, since $B = A$, $b = a = 52$.

   Also $c = a \div \sin 45° = 52\sqrt{2}$. **(Side plus angle)**

9. $c = 21.2$ and $b = 11.2$. Therefore

   $$a = \sqrt{c^2 - b^2} = \sqrt{324} = 18$$

   $$\sin A = \frac{a}{c} = \frac{18}{21.2} = .8491$$

   Using the calculator, $A = \text{Cos}^{-1}(11.2/21.2)(.8490) = 58.1092°$, $B = 90° – A = 31.8908°$. **(Two sides)**

10. $c = 80$ and $A = 52°$, thus $B = 90°–A = 38°$ and $a = 80 \sin 52° = 63.0409$,

    $b = 80 \cos 52° = 49.2529$. **(Side plus angle)**

11. c = 89 and a = 41. Therefore

$$b = \sqrt{c^2 - a^2} = \sqrt{6240} = 78.99$$

$$\sin A = \frac{a}{c} = \frac{41}{89} = .4607.$$

Using the calculator, A = Sin^{-1}(41/89) = 27.4306°, B = 90° − A = 62.5694°. **(Two sides)**

12. b = 60 and B = 30°, hence A = 90° − B = 60° and c = 60 ÷ sin 30° = 120.

Then a = 60 ÷ tan 30° = 60$\sqrt{3}$. **(Side plus angle)**

13. c = 132 and A = 16°, hence B = 90° − A = 74° and
a = 132 sin 16° = 36.3841, b = 132 cos 16° = 126.8865. **(Side plus angle)**

14. b = 31.3 and A = 67.8° so B = 90° − A = 22.2°.
Then c = 31.3 ÷ cos 67.8° = 82.8391
and a = 31.3 tan 67.8° = 76.6983. **(Side plus angle)**

15. a = 11.5 and b = 22.7. Therefore

$$c = \sqrt{a^2 + b^2} = \sqrt{647.54} = 25.4468$$

$$\tan A = \frac{a}{c} = \frac{11.5}{22.7} = .5066 \quad \text{and}$$

A = Tan^{-1}(11.5/22.7) = 26.8671°, B = 90° − A = 63.1329°. **(Two sides)**

16. c = 112.5 and B = 32.17°, thus A = 90° − 32.17° = 57.83° and
a = 112.5 sin 57.83° = 95.2281, b = 112.5 sin 32.17° = 59.8987. **(Side plus angle)**

17. a = 15.5 and B = 13.1°, so A = 90° − 13.1° = 76.9° and
c = 15.5 ÷ cos 13.1° = 15.5 sec 13.1° = 15.9142
b = 15.5 tan 13.1° = 3.6070. **(Side plus angle)**

18. b = 10 and B = 81°, therefore A = 90° − 81° = 9° and
c = 10 ÷ sin 81° = 10.1247, a = 10 tan 9° = 1.5838. **(Side plus angle)**

19. b = 10, and A = 81°, so B = 90° − 81° = 9° and
c = 10 ÷ sin 9° = 63.9245, a = 10 tan 81° = 63.1375. **(Side plus angle)**

20. c = 10 and A = 81°, so B = ,9° and
a = 10 sin 81° = 9.8769, b = 10 sin 9° = 1.5643. **(Side plus angle)**

21. d = 20 tan 68° = 49.5017m   **(Solving right triangles)**

Q. 21

22. $h = 46 \tan 77° = 199.2479m$ **(Solving right triangles)**

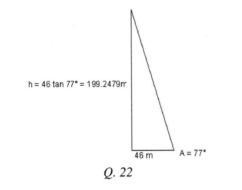

Q. 22

23. $d = 100 \tan 64° = 205.0304m$ **(Solving right triangles)**

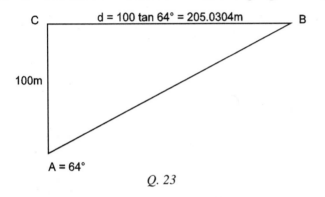

Q. 23

24. $h = d \tan 12.7° = (d - 400) \tan 13.1°$

$.2254 \, d = .2327 \, (d - 400)$

$.2254d = .2327d - 93.08$

$93.08 = .0073d$

$d = \dfrac{93.08}{.0073} = 12750.69$

$h = d \tan 12.7° = 2873.49m$

(Solving right triangles)

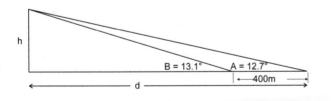

Q. 24

25. 11 mi = 58080 ft

sin A = 1300 ÷ 58080 = .0224

A = 1.28° **(Solving right triangles)**

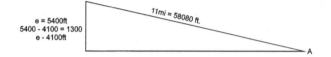

Q. 25

26. C = 180° − A − B = 55°

$$b = \frac{a \sin B}{\sin A} = 10.46$$

Since the triangle is isoceles we have
b = c = 10.46. **(Law of sines: ASA)**

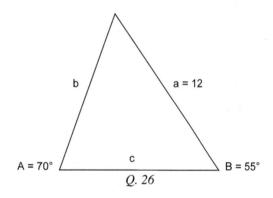

Q. 26

27. B = 180° − A − C = 96.5°

$$a = \frac{b \sin A}{\sin B} = 58.60$$

$$c = \frac{b \sin C}{\sin B} = 42.22 \quad \textbf{(Law of sines: ASA)}$$

Q. 27

28. $A = 180° - B - C = 42.5°$

$$a = \frac{c \sin A}{\sin C} = 167.72$$

$$b = \frac{c \sin B}{\sin C} = 237.41 \text{ (Law of sines: ASA)}$$

C = 30.5°

b

a

A

c = 126 *B* = 107°

Q. 28

29. $C = 180° - A - B = 120°$

$$a = \frac{c \sin A}{\sin C} = 44.77$$

$$b = \frac{c \sin B}{\sin C} = 32.99 \quad \text{(Law of sines: ASA)}$$

C

b *a*

A= 35° *B* = 25°

c = 67.6

Q. 29

30. $A = 180° - B - C = 42.8°$

$$b = \frac{a \sin B}{\sin A} = 1.44$$

$$c = \frac{a \sin C}{\sin A} = 1.56 \quad \text{(Law of sines: ASA)}$$

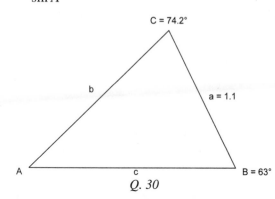

C = 74.2°

b

a = 1.1

A

c

B = 63°

Q. 30

31. $B = 180° - A - C = 85°$

$$a = \frac{b \sin A}{\sin B} = 22.63$$

$$c = \frac{b \sin C}{\sin B} = 36.69 \text{ (\textbf{Law of sines: ASA})}$$

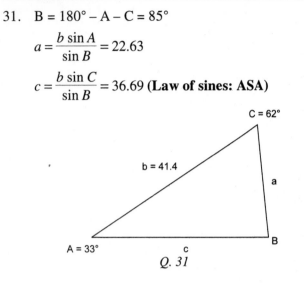

Q. 31

32.   Ambiguous case.

Since $A < 90°$ and $a > b$, there is only one solution.

$\sin B = (\,b/a\,) \sin A = .4821$

$B = \text{Sin}^{-1}(\,.4821) = 28.82°$

$C = 180° - A - B = 111.18°$

$c = a \sin C / \sin A = 29.01$ (**Law of sines: SSA**)

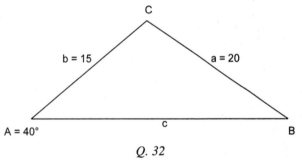

Q. 32

33.   Ambiguous case.

Since $A < 90°$ and $a > b$, there is only one solution.

$\sin B = (\,b/a\,) \sin A = .1590$

$B = \text{Sin}^{-1}(\,.1590) = 9.15°$

$C = 180° - A - B = 147.85°$

$c = a \sin C / \sin A = 73.95$ (**Law of sines: SSA**)

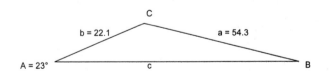

Q. 33

34. Ambiguous case.

    Since A < 90° and a > b, there is only one solution.

    sin B = (b/a) sin A = .7168

    B = Sin^{-1}(.7168) = 45.79°

    C = 180° – A – B = 9.21°

    c = a sin C / sin A = 7.82   **(Law of sines: SSA)**

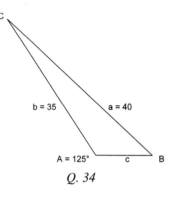

Q. 34

35. Ambiguous case.

    Since A < 90° and a > c, there is only one solution.

    sin C = (c/a) sin A = .7930

    C = Sin^{-1}(.7930) = 52.47°

    B = 180° – A – C = 64.53°

    b = a sin B / sin A = 10.13 **(Law of sines: SSA)**

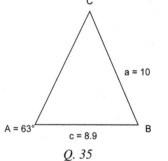

Q. 35

36. Ambiguous case.

    Since A < 90° and a > b, there may be only 0, 1, or 2 solutions.

    sin B = (b/a) sin A = 1.34

    Since sin B > 1, there is no solution.   **(Law of sines: SSA)**

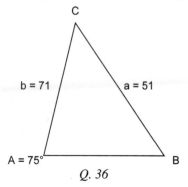

Q. 36

37. Ambiguous case.

    Since A > 90° and a > c, there is only one solution.

    sin C = (c/a) sin A = .6016

    C = Sin^{-1}(.6016) = 36.99°

    B = 180° – A – C = 7.01°

    b = a sin B / sin A = 10.10   **(Law of sines: SSA)**

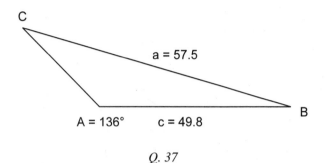

    Q. 37

38. Ambiguous case.

    Since A < 90° and a < b, there may be 0, 1, or 2 solutions.

    sin B = (b/a) sin A = .8074

    Since sin B < 1, there are two possible solutions.

    B = Sin^{-1}(.8074) = 53.85°

    B′ = 180° – B = 126.15°

    C = 180° – A – B = 89.15°

    C′ = 180° – A – B′ = 16.85°

    c = a sin C / sin A = 19.94 c′ = a sin C′ / sin A = 5.78   **(Law of sines: SSA)**

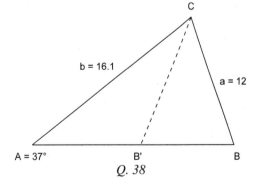

    Q. 38

39. Ambiguous case.

    Since A < 90° and a < b, there may be 0, 1, or 2 solutions.

    sin B = (b/a) sin A = 1.000076

    Since sin B > 1, there is no solution.

40. Ambiguous case.

   Since A < 90° and a < c, there may be 0, 1, or 2 solutions.

   $\sin C = (c/a) \sin A = .9290$

   Since $\sin C < 1$, there are two possible solutions.

   $C = \sin^{-1}(.9290) = 68.28°$

   $C' = 180° – C = 111.72°$

   $B = 180° – A – C = 53.72°$

   $B' = 180° – A – C' = 10.28°$

   $b = a \sin B / \sin A = 20.91$

   $b' = a \sin B' / \sin A = 4.63$ **(Law of sines: SSA)**

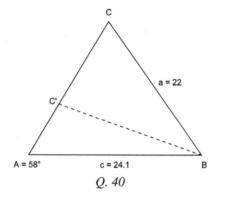

   Q. 40

41. Ambiguous case.

   Since A < 90° and a < b, there may be 0, 1, or 2 solutions.

   $\sin B = (b/a) \sin A = .8316$

   Since $\sin B < 1$, there are two solutions.

   $B = \sin^{-1}(.8316) = 56.27°$

   $B' = 180° – B = 123.73°$

   $C = 180° – A – B = 111.73°$

   $C' = 180° – A – B' = 44.27°$

   $c = a \sin C / \sin A = 31.28$

   $c' = a \sin C' / \sin A = 23.50$ **(Law of sines: SSA)**

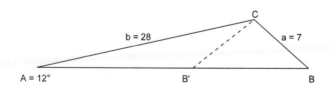

   Q. 41

42. Ambiguous case.

    Since A < 90° and a < b, there may be 0, 1, or 2 solutions.

    sin B = (b/a) sin A = 1.36.

    Since sin B > 1, there is no solution.

    No solution.   **(Law of sines: SSA)**

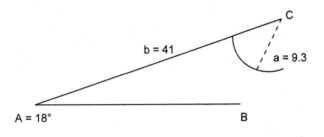

Q. 42

43. Ambiguous case.

    Since A < 90° and a < c, there may be 0, 1, or 2 solutions.

    sin C = (c/a) sin A = .9930

    Since sin C < 1, there are two possible solutions.

    C = Sin^{-1}(.9930) = 83.24°

    C′ = 180° – C = 96.76°

    B = 180° – A – C = 47.76°

    B′ = 180° – A – C′ = 34.24°

    b = a sin B / sin A = 93.19

    b′ = a sin B′/ sin A = 70.83   **(Law of sines: SSA)**

```
                        C
                       /|\
                      / | \
                   C'/  |  \
                   / \  |   \  a = 95
                  /   \ |    \
                 /     \|     \
                /       |\     \
               /        | \     \
          A = 49°       |  \     B
              A=49°  c = 125
```

Q. 43

44. Ambiguous case.

    Since A > 90° and a > c, there is 1 solution.

    sin C = (c/a) sin A = .8149

    C = Sin^{-1}(.8149) = 54.57°

    B = 180° – A – B = 13.43°

    b = a sin B / sin A = 10.55   **(Law of sines: SSA)**

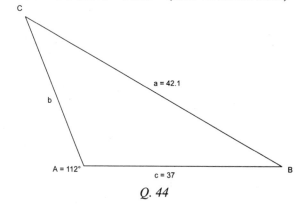

    Q. 44

45. Ambiguous case.

    Since A > 90° and a > b, there is 1 solution.

    sin B = (b/a) sin A = .2026

    B = Sin^{-1}(.2026) = 11.69°

    C = 180° – A – B = 6.31°

    c = a sin C / sin A = 2.17   **(Law of sines: SSA)**

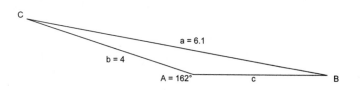

    Q. 45

46. $B = 180° - 44° - 57.3° = 78.7°$

$$\frac{\sin B}{850} = \frac{\sin 57.3°}{c}$$

$c = 850 \sin 57.3° \div \sin 78.7° = 729.42$ ft

$h = c \sin 44° = 506.70$ ft   **(Application law of sines)**

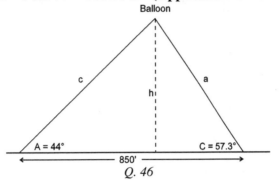

Q. 46

47. $A = 90° - 6° = 84°$ and $C = 180° - 22° - 84° = 74°$

$$\frac{\sin 22°}{H} = \frac{\sin 74°}{40}$$

$H = 40 \sin 22° \div \sin 74° = 15.59$ ft   **(Application law of sines)**

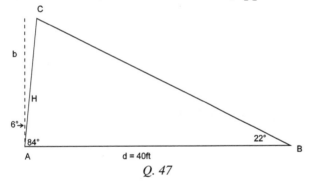

Q. 47

48. Angle $ABC = 180° - 85° = 95°$ and $C = 180° - 37° - 95° = 48°$

$$\frac{\sin 48°}{100} = \frac{\sin 37°}{L}$$

$L = 100 \sin 37° \div \sin 48° = 80.98$ ft   **(Application law of sines)**

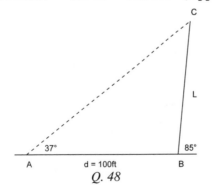

Q. 48

49. $C = 180° - 41° - 27° = 112°$

$$\frac{\sin 41°}{b} = \frac{\sin 112°}{800}$$

b = 800 sin 41° ÷ sin 112° = 566.07 ft

h = b sin 27° = 256.99 ft   **(Application law of sines)**

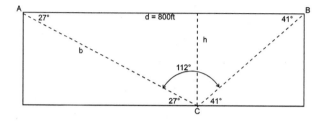

Q. 49

50. $a^2 = b^2 + c^2 - 2bc \cos A$

$a^2 = 29^2 + 17^2 - 2(29)(17) \cos 103° = 1351.80$

a = 36.77

$b^2 = a^2 + c^2 - 2ac \cos B$

841 = 1351.85 + 289 − 2(36.77)(17) cos *B*

cos *B* = .6398

$B = \cos^{-1}(.6397) = 50.22°$

C = 180° − B − A = 26.78°   **(Law of cosines: SAS)**

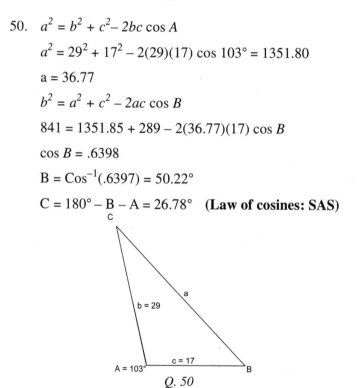

Q. 50

51. $a^2 = b^2 + c^2 - 2bc \cos A$

    $25 = 49 + 100 - 140 \cos A$

    $\cos A = .8857$

    $A = 27.66°$

    $b^2 = a^2 + c^2 - 2ac \cos B$

    $49 = 25 + 100 - 100 \cos B$

    $B = 40.54°$

    $C = 180° - B - A = 111.80$ **(Law of cosines: SSS)**

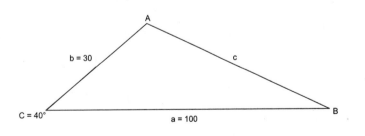

Q. 51

52. $c^2 = b^2 + a^2 - 2ab \cos C$

    $c^2 = 10,000 + 900 - 6000 \cos 40° = 6303.73$

    $c = 79.40$

    $a^2 = b^2 + c^2 - 2bc \cos A$

    $10,000 = 900 + 6303.73 - 4764 \cos A$

    $\cos A = -.5870$

    $A = 125.94°$ and $B = 180° - A - C = 14.06°$ **(Law of cosines: SAS)**

Q. 52

53. $a^2 = b^2 + c^2 - 2bc \cos A$

$49 = 576 + 676 - 2(24)(26) \cos A$

$\cos A = .9639$

$A = 15.43°$

$b^2 = a^2 + c^2 - 2ac \cos B$

$576 = 49 + 676 - 2(7)(26) \cos B$

$-149 = -364 \cos B$

$\cos B = .4093$

B = 65.84 and C = 180° − 15.43°− 65.84° = 98.73°   **(Law of cosines: SAS)**

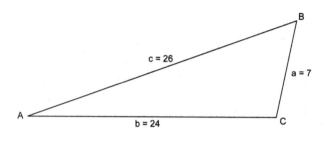

Q. 53

54. $c^2 = b^2 + a^2 - 2ab \cos C$

$c^2 = 70^2 + 82^2 - 2(70)(82) \cos 42 = 3092.70$

$c = 55.61$

$a^2 = b^2 + c^2 - 2bc \cos A$

$4900 = 6724 + 3092.7 - 2(82)(55.61) \cos A$

$\cos A = .5391$

A = 57.38° and B = 180° − A − C = 80.62°   **(Law of cosines: SAS)**

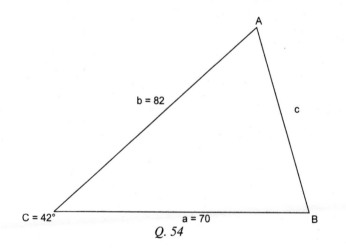

Q. 54

55. $a^2 = b^2 + c^2 - 2bc \cos A$

$a^2 = 91^2 + 67^2 - 2(91)(67) \cos 49° = 4770.02$

$a = 69.07$

$b^2 = a^2 + c^2 - 2ac \cos B$

$91^2 = 67^2 + 4770.02 - 2(67)(69.07) \cos B$

$\cos B = .1057$

B = 83.93° and C = 180° − B − A = 47.07°   **(Law of cosines: SSS)**

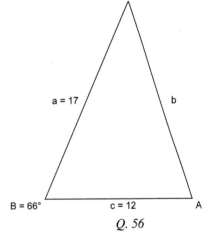

Q. 55

56. $b^2 = a^2 + c^2 - 2ac \cos B$

$b^2 = 17^2 + 12^2 - 2(17)(12) \cos 66° = 267.05$

$b = 16.34$ **(Law of cosines: SAS)**

Q. 56

57. The smallest angle is opposite the smallest side, b = 6.6.

$b^2 = a^2 + c^2 - 2ac \cos B$

$6.6^2 = 11.7^2 + 14.2^2 - 2(11.7)(14.2) \cos B$

$\cos B = .8877$

B = 27.41°   **(Law of cosines: SSS)**

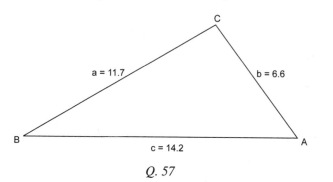

Q. 57

58. The largest angle is opposite the longest side, b = 40.2

$b^2 = a^2 + c^2 - 2ac \cos B$

$40.2^2 = 36.5^2 + 26.1^2 - 2(26.1)(36.5) \cos B$

$\cos B = .2086$

B = 77.96°   **(Law of cosines: SSS)**

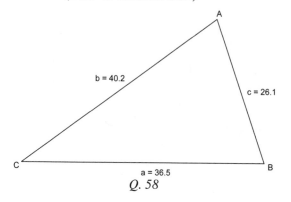

Q. 58

59. $a^2 = b^2 + c^2 - 2bc \cos A$

$a^2 = 11.2^2 + 48.2^2 - 2(11.2)(48.2) \cos 162° = 3475.52$

a = 58.95   **(Law of cosines: SAS)**

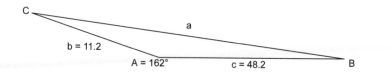

Q. 59

60. Since B = 90°, this is a right triangle.

    sin C = 11.9 ÷ 37 = .3216

    so C = 18.76°

    A = 90° − C = 71.24°

    and a = 37 sin A = 35.03   **(Solving right triangles)**

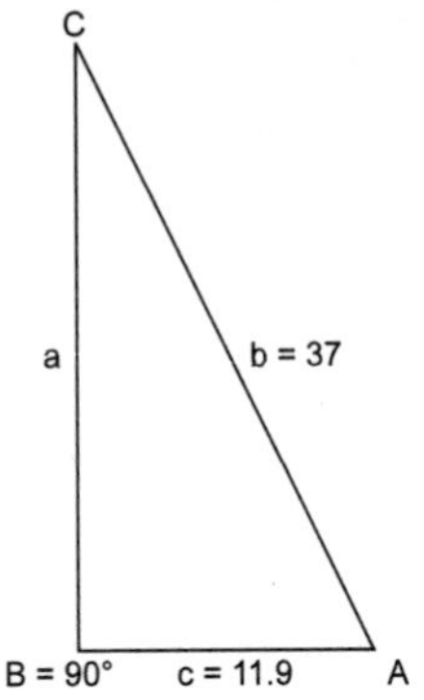

Q. 60

61. B = 180° − A − C = 115°

    $a = \dfrac{c \sin A}{\sin C} = 13.99$

    $b = \dfrac{c \sin B}{\sin C} = 41.02$ **(Law of sines: ASA)**

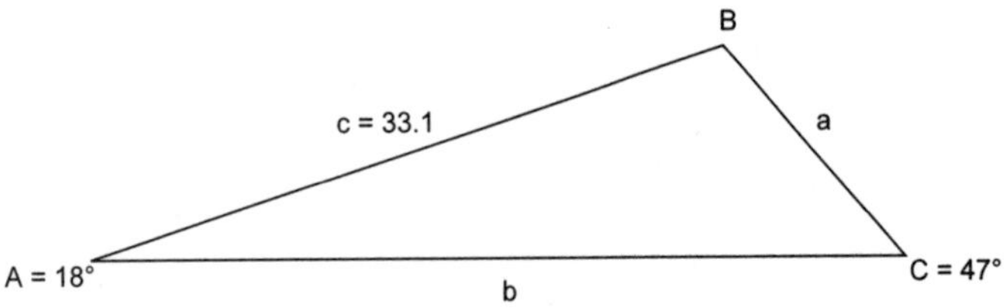

Q. 61

62. $a^2 = b^2 + c^2 - 2bc \cos A$

    $a^2 = 49 + 196 - 2(7)(14) \cos 58 = 141.14$

    $a = 11.88$

    $b^2 = a^2 + c^2 - 2ac \cos B$

    $49 = 141.14 + 196 - 2(11.88)(14) \cos B$

    $\cos B = .8662$

    B = 29.98°, C = 180° − B − A = 92.02°   **(Law of cosines: SAS)**

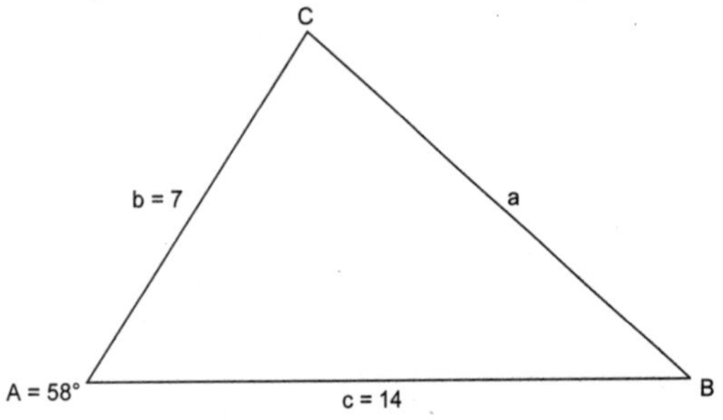

Q. 62

63. $c^2 = b^2 + a^2 - 2ab \cos C$

$c^2 = (30.3)^2 + 17^2 - 2(17)(30.3) \cos 71° = 871.69$

$c = 29.52$

$a^2 = b^2 + c^2 - 2bc \cos A$

$17^2 = 30.3^2 + 871.69 - 2(30.3)(29.52) \cos A$

$\cos A = .8389$

$A = 32.97°$ $B = 180° - A - C = 76.03°$ **(Law of cosines: SAS)**

Q. 63

64. $\dfrac{\sin A}{a} = \dfrac{\sin B}{b} = \dfrac{\sin C}{c}$

$\sin B = \dfrac{b \sin A}{a} = \dfrac{16.36}{20} = .8181$

$B = 54.90°$ and $C = 180° - B - A = 86.10°$

$c = \dfrac{a \sin C}{\sin A} = 31.71$

or:

$B' = 180° - B = 125.10°$ and $C' = 180° - B' - A = 15.90°$

$c' = \dfrac{a \sin C'}{\sin A} = 8.71$ **(Law of Sines: SSA, ambiguous case)**

Q. 64

Grade Yourself

Circle the numbers of the questions you missed, then fill in the total incorrect for each topic. If you answered more than three questions incorrectly, you need to focus on that topic. (If a topic has less than three questions and you had at least one wrong, we suggest you study that topic also. Read your textbook, a review book, or ask your teacher for help.)

Subject: Solving Triangles

Topic	Question Numbers	Number Incorrect
Properties of triangles	1, 2, 3, 4, 5	
Side plus angle	6, 8, 10, 12, 13, 14, 16, 17, 18, 19, 20	
Two sides	7, 9, 11, 15	
Solving right triangles	21, 22, 23, 24, 25, 60	
Law of sines: SAS	26, 27, 28, 29, 30, 31	
Law of sines: ASA	61	
Law of sines: SSA, ambiguous case	32, 33, 34, 35, 36, 37, 38, 39, 40, 41, 42, 43, 44, 45, 64	
Application law of sines	46, 47, 48, 49	
Law of cosines: SAS	50, 52, 54, 55, 56, 59, 62, 63	
Law of cosines: SSS	51, 53, 57, 58	

Trigonometric Identities

6

Brief Yourself

A trigonometric identity is an expression that is true for all values of the angles involved. On the other hand, a trigonometric equation is one that is true for only certain values of the angles in the equation. In some cases all possible solutions are desired, while in other cases we are only interested in solutions in a certain range, say between 0 and 2π. For example, $\sin x = 1/2$ is satisfied by $x = \pi/6$ and by $x = 5\pi/6$; these are the only solutions in the range from 0 to 2π. However, the equation is also satisfied by $x = \pi/6 + 2n\pi$ and by $x = 5\pi/6 + 2n\pi$ for all integer values of n. Thus, the set of all possible solutions contains infinitely many entries.

Some equations can be solved by simply using an appropriate inverse trigonometric function. More complicated equations may require factoring or the application of trigonometric identities. When factoring, recall that an equation that asserts that the product of several factors equals zero is satisfied if any one of the factors equals zero. The set of solutions to such an equation may be obtained by setting each of the separate factors equal to zero and solving.

Summary of Identities

1. $\tan \theta = \dfrac{\sin \theta}{\cos \theta}$ $\qquad$ $\cot \theta = \dfrac{\cos \theta}{\sin \theta}$

2. $\csc \theta = \dfrac{1}{\sin \theta}$ $\qquad$ $\sec \theta = \dfrac{1}{\cos \theta}$ $\qquad$ $\cot \theta = \dfrac{1}{\tan \theta}$

3. $\sin^2 \theta + \cos^2 \theta = 1$ $\quad$ $\tan^2 + 1 = \sec^2 \theta$ $\qquad$ $1 + \cot^2 \theta = \csc^2 \theta$

4. $\sin(-\theta) = -\sin(\theta)$ $\quad$ $\cos(-\theta) = \cos(\theta)$ $\qquad$ $\tan(-\theta) = -\tan(\theta)$

5. $\sin(\alpha \pm \beta) = \sin \alpha \cos \beta \pm \sin \beta \cos \alpha$

    $\cos(\alpha \pm \beta) = \cos \alpha \cos \beta \mp \beta \sin \alpha$

    $\tan(\alpha \pm \beta) = \dfrac{\tan \alpha \pm \tan \beta}{1 \mp \tan \alpha \tan \beta}$

6. $\sin\left(\dfrac{\pi}{2} - \theta\right) = \cos \theta$ $\quad$ $\cos\left(\dfrac{\pi}{2} - \theta\right) = \sin \theta$ $\qquad$ $\tan\left(\dfrac{\pi}{2} - \theta\right) = \cot \theta$

7. $\sin 2\theta = 2 \sin \theta \cos \theta$

    $\cos 2\theta = \cos^2 \theta - \sin^2 \theta = 2 \cos^2 \theta - 1 = 1 - 2 \sin^2 \theta$

    $\tan 2\theta = \dfrac{2 \tan \theta}{1 - \tan^2 \theta}$

8. $\sin(\theta/2) = \pm \sqrt{\dfrac{1 - \cos \theta}{2}}$ $\cos(\theta/2) = \pm \sqrt{\dfrac{1 + \cos \theta}{2}}$

    $\tan(\theta/2) = \dfrac{1 - \cos \theta}{\sin \theta} = \dfrac{\sin \theta}{1 + \cos \theta}$

Test Yourself

Verify each of the following identities:

1. $\cos x(\tan x + \cot x) = \csc x$

2. $\csc x \cos x = \cot x$

3. $\csc x - \cos x \cot x = \sin x$

4. $1/(\sec x + \tan x) + 1/(\sec x - \tan x) = 2 \sec x$

5. $\cot x \tan^2 x = \tan x$

Evaluate the required trigonometric function using identities 1, 2, or 3:

6. If $\sin x = 1/5$, then find $\csc x$.

7. If $\csc x = \sqrt{2}$, and x is an angle in quadrant I, then find $\cot x$.

8. If $\sec x = 2$ and x is a quadrant IV angle, then find $\csc x$.

9. If $\cos x = -1/2$ and x is a quadrant II angle, then find $\tan x$.

10. If $\tan x = -6$ and x is a quadrant IV angle, then find $\sec x$.

Verify the following identities:

11. $\cos^2 x - \sin^2 x = 2 \cos^2 x - 1$

12. $\sec x - \tan x = \cos x/(1 + \sin x)$

13. $\tan x + \cot x = \sec x \csc x$

14. $(1 + \tan^2 x) \cos^2 x = 1$

15. $\sin^3 x = \sin x - \sin x \cos^2 x$

16. $\sin x /(1 + \cos x) + (1 + \cos x)/\sin x = 2 \csc x$

17. $(1 - \cos x)/(1 + \cos x) = (\csc x - \cot x)^2$

18. $1/(\sec x - \tan x) = \sec x + \tan x$

19. $\sin x \cos x /(\cos^2 x - \sin^2 x) = \tan x /(1 - \tan^2 x)$

20. $\cos^6 x = 1 - 3\sin^2 x + 3\sin^4 x - \sin^6 x$

Evaluate the following expressions using identities 4, 5, or 6:

21. $\tan(-\pi/3)$

22. $\cos(-\pi/4)$

23. $\cos \pi + \cos(-\pi)$

24. $-\sin(-\pi/3)$

25. $\cos(\pi/2 - 2\pi)$

26. $\tan(\pi/2 - \pi/3) - \sin(\pi/2 - \pi/3)$

27. $\cos(\pi/12)$
    (Hint: $\cos(\pi/12) = \cos(\pi/3 - \pi/4)$)

28. $\sin 75°$
    (Hint: $\sin 75° = \sin(30° + 45°)$)

29. $\sin 105°$

30. $\cos 15°$

31. $\sin(\operatorname{Sin}^{-1}(1/2) + \operatorname{Cos}^{-1}0)$

32. $\cos(\operatorname{Tan}^{-1}(4/3) + \operatorname{Cos}^{-1}(5/13))$

33. $\tan(\operatorname{Sin}^{-1}(1/2) - \operatorname{Cos}^{-1}(1/2))$

34. $\cos(\operatorname{Cos}^{-1}(\sqrt{3}/2) + \operatorname{Sin}^{-1}(\sqrt{3}/2))$

35. $\tan(\operatorname{Cot}^{-1}1 - \operatorname{Sin}^{-1}(\sqrt{2}/2))$

Verify the following identities:

36. $\sin(x+y) / \sin(x - y) = (\tan x + \tan y)/(\tan x - \tan y)$

37. $\tan(2\pi - x) = -\tan x$

38. $\cos(\pi \pm x) = -\cos x$

39. $\cos(x - y)\cos(x+y) = \cos^2 x - \sin^2 y$

40. $\sin(x + \pi/4) = \cos(\pi/4 - x)$

41. $\cos(x+y) / (\cos x \cos y) = 1 - \tan x \tan y$

42. $\cos(\pi/2 + x - y) = \sin(y - x)$

43. $\sin(x+y)\sin(x - y) = \sin^2 x - \sin^2 y$

44. $\cos(x+y) + \cos(x - y) = 2\cos x \cos y$

45. $\sin(x+y) + \sin(x - y) = 2\sin x \cos y$

Evaluate sin 2x, cos 2x, tan 2x, and find the quadrant of 2x if:

46. $\sin x = 4/5$ and x is in quadrant I

47. $\cos x = -\sqrt{3}/2$ and x is in quadrant III

48. $\cos x = -5/13$ and x is in quadrant II

49. $\sin x = 3/5$ and x is in quadrant II

50. $\tan x = -3/4$ and x is in quadrant IV

If $\sin x = .6157$ and x is in quadrant I then evaluate:

51. $\sin 2x$

52. $\cos 2x$

53. $\sin 4x$

54. $\sin(x/2)$

55. $\cos(x/2)$

Verify the following identities:

56. $(1 + \cos 2x)/(1 - \cos 2x) = \cot^2 x$

57. $\sin 2x / (1 + \cos 2x) = \tan x$

58. $(\cos^4 x - \sin^4 x)/(1 - \tan^4 x) = \cos^4 x$

59. $\sec^2 x / (2 - \sec^2 x) = \sec 2x$

60. $1 - \cos 5x \cos 3x - \sin 5x \sin 3x = 2\sin^2 x$

Simplify the following expressions:

61. $\cos^4 x - \sin^4 x$

62. $(\sin x - \cos x)^2 + \sin 2x$

63. $2\cos^2(x/2) - \cos x$

64. $2\sin^2(x/2) + \cos x$

65. $\sin 2x / \cos x$

✔ Check Yourself

The identities will be verified by bringing the left side of the identity into agreement with the right side. There are often many different ways of proving an identity

1. $\cos x(\tan x + \cot x) = \sin x + \dfrac{\cos^2 x}{\sin x} = \sin x + \dfrac{1 - \sin^2 x}{\sin x}$

   $= \sin x + \dfrac{1}{\sin x} - \dfrac{\sin^2 x}{\sin x} = \sin x + \dfrac{1}{\sin x} - \sin x = \dfrac{1}{\sin x} = \csc x$ **(Application of identities 1, 2, 3)**

2. $\csc x \cos x = \dfrac{1}{\sin x} \cos x = \cot x$

3. $\csc x - \cos x \cot x = \dfrac{1}{\sin x} - \dfrac{\cos^2 x}{\sin x} = \dfrac{1 - \cos^2 x}{\sin x} = \dfrac{\sin^2 x}{\sin x} = \sin x$ **(Application of identities 1, 2, 3)**

4. $\dfrac{1}{\sec x + \tan x} + \dfrac{1}{\sec x - \tan x} = \dfrac{(\sec x - \tan x) + (\sec x + \tan x)}{(\sec x + \tan x)(\sec x - \tan x)} = \dfrac{2 \sec x}{\sec^2 x - \tan^2 x} = \dfrac{2 \sec x}{1} = 2 \sec x$

   (Application of identities 1, 2, 3)

5. $\cot x \tan^2 x = (\cot x \tan x)\tan x = \tan x$ **(Application of identities 1, 2, 3)**

6. If $\sin x = 1/5$, then $\csc x = 1/\sin x = 5$.   **(Application of identities 1, 2, 3)**

7. If $\csc x = \sqrt{2}$ and x is in quadrant I, then $\cot^2 x = \csc^2 x - 1 = 2 - 1 = 1$ and $\cot x = 1$, since the cotangent is positive in the first quadrant.   **(Application of identities 1, 2, 3)**

8. If $\sec x = 2$ then $\cos x = 1/2$ and $\sin^2 x = 1 - \cos^2 x = 3/4$. For x in the fourth quadrant $\sin x$ is negative, hence $\sin x = -\sqrt{3}/2$. Finally, $\csc x = -2/\sqrt{3}$.   **(Application of identities 1, 2, 3)**

9. If $\cos x = -1/2$, then $\sec x = -2$ and $\sec^2 x = 4$. Then $\tan^2 x = \sec^2 x - 1 = 4 - 1 = 3$, and for x in quadrant II $\tan x$ is negative, so, $\tan x = -\sqrt{3}$.   **(Application of identities 1, 2, 3)**

10. If $\tan x = -6$ then $\sec^2 x = 36 + 1$, and for x in quadrant IV $\sec x$ is positive, so $\sec x = \sqrt{37}$.   **(Application of identities 1, 2, 3)**

11. $\cos^2 x - \sin^2 x = \cos^2 x - (1 - \cos^2 x) = 2\cos^2 x - 1$ **(Application of identities 1, 2, 3)**

12. $\sec x - \tan x = \dfrac{1 - \sin x}{\cos x} = \dfrac{1 - \sin^2 x}{\cos x(1 + \sin x)} = \dfrac{\cos^2 x}{\cos x(1 + \sin x)} = \dfrac{\cos x}{1 + \sin x}$ **(Application of identities 1, 2, 3)**

13. $\tan x + \cot x = \dfrac{\cos x}{\sin x} + \dfrac{\sin x}{\cos x} = \dfrac{\cos^2 x + \sin^2 x}{\sin x \cos x} = \dfrac{1}{\sin x \cos x} = \csc x \sec x$ **(Application of identities 1, 2, 3)**

14. $(1 + \tan^2 x)\cos^2 x = \sec^2 x \cos^2 x = 1$ **(Application of identities 1, 2, 3)**

15. $\sin^3 x = \sin x \sin^2 x = \sin x(1 - \cos^2 x) = \sin x - \sin x \cos^2 x$ **(Application of identities 1, 2, 3)**

16. $\dfrac{\sin x}{1+\cos x}+\dfrac{1+\cos x}{\sin x}=\dfrac{\sin x(1-\cos x)}{1-\cos^2 x}+\dfrac{1+\cos x}{\sin x}=\dfrac{1-\cos x}{\sin x}+\dfrac{1+\cos x}{\sin x}=\dfrac{2}{\sin x}=2\csc x$ **(Application of identities 1, 2, 3)**

17. $\dfrac{1-\cos x}{1+\cos x}=\dfrac{(1-\cos x)^2}{1-\cos^2 x}=\dfrac{(1-\cos x)^2}{\sin^2 x}=\left(\dfrac{1-\cos x}{\sin x}\right)^2=\left(\dfrac{1}{\sin x}-\dfrac{\cos x}{\sin x}\right)^2=(\csc x-\cot x)^2$ **(Application of identities 1, 2, 3)**

18. $\dfrac{1}{\sec x-\tan x}=\dfrac{\sec x+\tan x}{\sec^2 x-\tan^2 x}=\dfrac{\sec x+\tan x}{1}$ **(Application of identities 1, 2, 3)**

19. $\dfrac{\sin x\cos x}{\cos^2 x-\sin^2 x}=\dfrac{\sin x/\cos x}{1-\sin^2 x/\cos^2 x}=\dfrac{\tan x}{1-\tan^2 x}$ **(Application of identities 1, 2, 3)**

20. $\cos^6 x=(\cos^2 x)^3=(1-\sin^2 x)^3=1-3\sin^2 x+3\sin^4 x-\sin^6 x$ **(Application of identities 1, 2, 3)**

21. $\tan(-\pi/3)=-\tan(\pi/3)=-\sqrt{3}$ **(Application of identities 4, 5, 6)**

22. $\cos(-\pi/4)=\cos(\pi/4)=\sqrt{2}/2$ **(Application of identities 4, 5, 6)**

23. $\cos\pi+\cos(-\pi)=2\cos\pi=-2$ **(Application of identities 4, 5, 6)**

24. $-\sin(-\pi/3)=\sin(\pi/3)=\sqrt{3}/2$ **(Application of identities 4, 5, 6)**

25. $\cos(\pi/2-2\pi)=\cos(\pi/2)\cos 2\pi+\sin(\pi/2)\sin 2\pi=0$ **(Application of identities 4, 5, 6)**

26. $\tan(\pi/2-\pi/3)-\sin(\pi/2-\pi/3)=\cot(\pi/3)-\cos(\pi/3)=(1/\sqrt{3})-1/2$ Note: the formula for $\tan(x-y)$ does not apply since $\tan(\pi/2)$ is undefined. **(Application of identities 4, 5, 6)**

27. $\cos(\pi/12)=\cos(\pi/3-\pi/4)=\cos(\pi/3)\cos(\pi/4)+\sin(\pi/3)\sin(\pi/4)=(1+\sqrt{3})/(2\sqrt{2})=(\sqrt{2}+\sqrt{6})/4$ **(Application of identities 4, 5, 6)**

28. $\sin 75°=\sin(30°+45°)=\sin 30°\cos 45°+\sin 45°\cos 30°=(1+\sqrt{3})/(2\sqrt{2})=(\sqrt{2}+\sqrt{6})/4$ **(Application of identities 4, 5, 6)**

29. $\sin 105°=\sin(45°+60°)=\sin 45°\cos 60°+\sin 60°\cos 45°=(1+\sqrt{3})/(2\sqrt{2})=(\sqrt{2}+\sqrt{6})/4$ **(Application of identities 4, 5, 6)**

30. $\cos 15°=\cos(45°-30°)=\cos 45°\cos 30°+\sin 45°\sin 30°=(1+\sqrt{3})/(2\sqrt{2})=(\sqrt{2}+\sqrt{6})/4$ **(Application of identities 4, 5, 6)**

31. $\sin(\mathrm{Sin}^{-1}(1/2)+\mathrm{Cos}^{-1}0)=\sin(\mathrm{Sin}^{-1}(1/2))\cos(\mathrm{Cos}^{-1}0)+\cos(\mathrm{Sin}^{-1}(1/2))\sin(\mathrm{Cos}^{-1}0)$

$=\sin x\cos y+\cos x\sin y=\sqrt{3}/2$

Note: since $\sin x=1/2$, $x=\pi/6$ and $\cos x=\sqrt{3}/2$,

$\cos y=0$, $y=\pi/2$ and $\sin y=1$.

Alternate solution: $\sin(\mathrm{Sin}^{-1}(1/2)+\mathrm{Cos}^{-1}0)=\sin(\pi/6+\pi/2)=\sin 2\pi/3=\sqrt{3}/2$ **(Application of identities 4, 5, 6)**

32. $\cos(\text{Tan}^{-1}(4/3) + \text{Cos}^{-1}(5/13)) = \cos x \cos y - \sin x \sin y$

    Note: $\tan x = 4/3$ implies $\cos x = 3/5$ and $\sin x = 4/5$

      $\cos y = 5/13$ implies $\sin y = 12/13$

    $\cos(\text{Tan}^{-1}(4/3) + \text{Cos}^{-1}(5/13)) = (3/5)(5/13) - (4/5)(12/13) = -33/65$ **(Application of identities 4, 5, 6)**

33. $\tan(\text{Sin}^{-1}(1/2) - \text{Cos}^{-1}(1/2)) = (\tan x - \tan y)/(1 + \tan x \tan y)$

    Note: $\sin x = 1/2$ implies $\tan x = 1/\sqrt{3}$

      $\cos y = 1/2$ implies $\tan y = \sqrt{3}$

    $\tan(\text{Sin}^{-1}(1/2) - \text{Cos}^{-1}(1/2)) = ((1/\sqrt{3}) - \sqrt{3})/2 = (\sqrt{3}/3 - 3\sqrt{3})/2 = -(2\sqrt{3}/3)$ **(Application of identities 4, 5, 6)**

34. $\cos(\text{Cos}^{-1}(\sqrt{3}/2) + \text{Sin}^{-1}(\sqrt{3}/2)) = 0$ or $\text{Cos}(\pi/6 = \pi/3) = \cos(\pi/2) = 0$

    Since $\text{Cos}^{-1}(\sqrt{3}/2) = \pi/6$ and $\text{Sin}^{-1}\sqrt{3}/2) = \pi/3$,

    and therefore $\text{Cos}^{-1}(\sqrt{3}/2) + \text{Sin}^{-1}(\sqrt{3}/2) = \pi//2$ **(Application of identities 4, 5, 6)**

35. $\tan(\text{Cot}^{-1}1 - \text{Sin}^{-1}(\sqrt{2}/2)) = 0$

    Since $\cot x = 1$, $\tan x = 1$, and $\sin y = \sqrt{2}/2$, so $\tan y = 1$.   **(Application of identities 4, 5, 6)**

36. $\dfrac{\sin(x+y)}{\sin(x-y)} = \dfrac{\sin x \cos y + \sin y \cos x}{\sin x \cos y - \sin y \cos x}$

    Divide all terms by $\cos x \cos y$

    $= \dfrac{(\sin x / \cos x) + (\sin y / \cos y)}{(\sin x / \cos x) - (\sin y / \cos y)} = \dfrac{\tan x + \tan y}{\tan x - \tan y}$ **(Application of identities 4, 5, 6)**

37. $\tan(2\pi - x) = \dfrac{\tan 2\pi - \tan x}{1 + \tan 2\pi \tan x} = \dfrac{-\tan x}{1}$ **(Application of identities 4, 5, 6)**

38. $\cos(\pi \pm x) = \cos \pi \cos x \mp \sin \pi \sin x = -\cos x$ **(Application of identities 4, 5, 6)**

39. $\cos(x - y)\cos(x + y) = (\cos x \cos y + \sin x \sin y)(\cos x \cos y - \sin x \sin y)$

    $= \cos^2 x \cos^2 y - \sin^2 x \sin^2 y$

    $= \cos^2 x(1 - \sin^2 y) - (1 - \cos^2 x)\sin^2 y$

    $= \cos^2 x - \cos^2 x \sin^2 y - \sin^2 y + \cos^2 x \sin^2 y$

    $= \cos^2 x - \sin^2 y$ **(Application of identities 4, 5, 6)**

40. $\sin(x + \pi/4) = \sin x \cos \pi/4 + \sin \pi/4 \cos x$

    $= \sin x \sin \pi/4 + \cos \pi/4 \cos x = \cos(\pi/4 - x)$ **(Application of identities 4, 5, 6)**

41. $\dfrac{\cos(x+y)}{\cos x \cos y} = \dfrac{\cos x \cos y - \sin x \sin y}{\cos x \cos y} = 1 - \dfrac{\sin x \sin y}{\cos x \cos y} = 1 - \tan x \tan y$ **(Application of identities 4, 5, 6)**

42. $\cos(\pi/2 + x - y) = \cos(\pi/2)\cos(x - y) - \sin(\pi/2)\sin(x - y)$

    $= -\sin(\pi/2)\sin(x - y) = -\sin(x - y) = \sin(y - x)$ **(Application of identities 4, 5, 6)**

43. $\sin(x+y)\sin(x-y) = (\sin x \cos y + \sin y \cos x)(\sin x \cos y - \sin y \cos x)$

$$= \sin^2 x \cos^2 y - \sin^2 y \cos^2 x$$

$$= \sin^2 x (1 - \sin^2 y) - (1 - \sin^2 x) \sin^2 y$$

$$= \sin^2 x - \sin^2 y \quad \textbf{(Application of identities 4, 5, 6)}$$

44. $\cos(x+y) + \cos(x-y)$

$= \cos x \cos y - \sin x \sin y + \cos x \cos y + \sin x \sin y$

$= 2 \cos x \cos y \quad \textbf{(Application of identities 4, 5, 6)}$

45. $\sin(x+y) + \sin(x-y)$

$= \sin x \cos y + \sin y \cos x + \sin x \cos y - \sin y \cos x$

$= 2 \sin x \cos y \quad \textbf{(Application of identities 4, 5, 6)}$

46. If $\sin x = 4/5$ and x is in quadrant I then $\cos x = 3/5$ and $\tan x = 4/3$

Then $\sin 2x = 2 \sin x \cos x = 24/25$

$\cos 2x = \cos^2 x - \sin^2 x = -7/25$

$\tan 2x = 2 \tan x/(1 - \tan^2 x) = -24/7$. Note that $\tan 2x$ can also be found by using $\sin 2x / \cos 2x = \tan 2x$.

$2x$ lies in quadrant II since $\sin 2x$ is positive but the cosine is negative. **(Application of identities 7)**

47. If $\cos x = -\sqrt{3}/2$ and x is in quadrant III, then $\sin x = -1/2$ and $\tan x = 1/\sqrt{3}$.

Then, $\sin 2x = 2 \sin x \cos x = 2\left(-\dfrac{1}{2}\right)\left(-\dfrac{\sqrt{3}}{2}\right) = \sqrt{3}/2$

$\cos 2x = \cos^2 x - \sin^2 x = (-\sqrt{3}/2)^2 - (-1/2)^2 = 3/4 - 1/4 = 1/2$

$\tan 2x = \sin 2x / \cos 2x = (\sqrt{3}/2)/(1/2) = \sqrt{3}$

and $2x$ lies in quadrant I. **(Application of identities 7)**

48. If $\cos x = -5/13$ and x is in quadrant II then $\sin x = 12/13$ and $\tan x = -12/5$. Then

$\sin 2x = -120/169$, $\cos 2x = -119/169$, $\tan 2x = 120/119$ and $2x$ lies in quadrant III **(Application of identities 7)**

49. If $\sin x = 3/5$ and x is in quadrant II then $\cos x = -4/5$ and $\tan x = -3/4$ .Then

$\sin 2x = -24/25$, $\cos 2x = 7/25$, $\tan 2x = -24/7$ and $2x$ lies in quadrant IV. **(Application of identities 7)**

50. If $\tan x = -3/4$ and x is in quadrant IV then $\sin x = -3/5$ and $\cos x = 4/5$. Then

$\sin 2x = -24/25$, $\cos 2x = 7/25$, $\tan 2x = -24/7$ and $2x$ lies in quadrant IV. **(Application of identities 7)**

51. $\cos x = +\sqrt{1 - (.6157)^2} = \sqrt{.6209} = 0.7880$

$\sin 2x = 2(.6157)(.7880) = .9703$ **(Application of identities 7, 8)**

52. $\cos 2x = (.7880)^2 - (.6157)^2 = .2419$

or $\cos 2x = \sqrt{1 - \sin^2 2x} = \sqrt{1 - (.9703)^2}$ **(Application of identities 7, 8)**

53. $\sin 4x = 2 \sin 2x \cos 2x = 2(.9703)(.2419) = .4694$ **(Application of identities 7, 8)**

54. $\sin(x/2) = +\sqrt{\dfrac{1 - \cos x}{2}} = \sqrt{\dfrac{1 - .7880}{2}} = 0.3256$ (+ because x is in quad I) **(Application of identities 7, 8)**

55. $\cos(x/2) = +\sqrt{\dfrac{1+\cos x}{2}} = \sqrt{\dfrac{1.7880}{2}} = 0.9455$ (+ because x is in quad I)

    or $\cos(x/2) = \sqrt{1 - \sin^2(x/2)} = \sqrt{.89398}$ **(Application of identities 7, 8)**

56. $\dfrac{1+\cos 2x}{1-\cos 2x} = \dfrac{1+(2\cos^2 x -1)}{1-(1-2\sin^2 x)} = \dfrac{2\cos^2 x}{2\sin^2 x} = \cot^2 x$ **(Application of identities 5, 6, 7, 8)**

57. $\dfrac{\sin 2x}{1+\cos 2x} = \dfrac{2\sin x \cos x}{2\cos^2 x} = \dfrac{\sin x}{\cos x} = \tan x$ **(Application of identities 5, 6, 7, 8)**

58. $\dfrac{\cos^4 x - \sin^4 x}{1-\tan^4 x} = \dfrac{(\cos^4 x - \sin^4 x)\cos^4 x}{(1-(\sin^4 x/\cos^4 x))\cos^4 x} = \dfrac{(\cos^4 x - \sin^4 x)\cos^4 x}{\cos^4 x - \sin^4 x} = \cos^4 x \dfrac{\cos^4 - \sin^4 x}{\cos^4 x - \sin^4 x} = \cos^4 x$
    (Application of identities 5, 6, 7, 8)

59. $\dfrac{\sec^2 x}{2-\sec^2 x} = \dfrac{1}{\cos^2 x(2-\sec^2 x)} = \dfrac{1}{2\cos^2 x - 1} = \dfrac{1}{\cos 2x} = \sec 2x$ **(Application of identities 5, 6, 7, 8)**

60. $1 - \cos 5x \cos 3x - \sin 5x \sin 3x = 1 - (\cos 5x \cos 3x + \sin 5x \sin 3x) = 1 - \cos 2x = 2\sin^2 x$
    (Application of identities 5, 6, 7, 8)

61. $\cos^4 x - \sin^4 x = (\cos^2 x - \sin^2 x)(\cos^2 x + \sin^2 x) = (\cos^2 x - \sin^2) \cdot 1 = \cos 2x$ **(Application of identities)**

62. $(\sin x - \cos x)^2 + \sin 2x = \sin^2 x - 2\sin x \cos x + \cos^2 x + 2\sin x \cos x = \sin^2 x + \cos^2 x = 1$ **(Application of identities)**

63. $2\cos^2(x/2) - \cos x = (1+\cos x) - \cos x = 1$ **(Application of identities)**

64. $2\sin^2(x/2) + \cos x = (1-\cos x) + \cos x = 1$ **(Application of identities)**

65. $\sin 2x / \cos x = (2\sin x \cos x)/\cos x = 2\sin x$ **(Application of identities)**

Grade Yourself

Circle the numbers of the questions you missed, then fill in the total incorrect for each topic. If you answered more than three questions incorrectly, you need to focus on that topic. (If a topic has less than three questions and you had at least one wrong, we suggest you study that topic also. Read your textbook, a review book, or ask your teacher for help.)

Subject: Trigonometric Identities

Topic	Question Numbers	Number Incorrect
Application of identities 1, 2, 3	1, 2, 3, 4, 5, 6, 7, 8, 9, 10, 11, 12, 13, 14, 15, 16, 17, 18, 19, 20	
Application of identities 4, 5, 6	21, 22, 23, 24, 25, 26, 27, 28, 29, 30, 31, 32, 33, 34, 35, 36, 37, 38, 39, 40, 41, 42, 43, 44, 45	
Application of identities 7	46, 47, 48, 49, 50	
Application of identities 7, 8	51, 52, 53, 54, 55	
Application of identities 5, 6, 7, 8	56, 57, 58, 59, 60	
Application of identities	61, 62, 63, 64, 65	

Trigonometric Equations

Brief Yourself

A trigonometric identity is an expression that is true for all values of the angles involved, whereas a trigonometric equation is one that is true for only certain values of the angles in the equation. In some cases all possible solutions are desired, while in other cases we are only interested in solutions in a certain range, say between 0 and 2π. For example, $\sin x = 1/2$ is satisfied by $x = \pi/6$ and by $x = 5\pi/6$; these are the only solutions in the range from 0 to 2π. However, the equation is also satisfied by $x = \pi/6 + 2n\pi$ and by $x = 5\pi/6 + 2n\pi$ for all integer values of n. Thus, the set of all possible solutions contains infinitely many entries.

Some equations can be solved by simply using an appropriate inverse trigonometric function. More complicated equations may require factoring or the application of trigonometric identities. When factoring, recall that an equation that asserts that the product of several factors equals zero is satisfied if any one of the factors equals zero. The set of solutions to such an equation may be obtained by setting each of the separate factors equal to zero and solving.

Test Yourself

Find all solutions of the following equations (without using a calculator):

1. $\cos x = 1/\sqrt{2}$

2. $\tan x = 1$

3. $\sin x = -1/2$

4. $\cos x = -1/2$

5. $\sin x = -1$

6. $\tan x = 0$

7. $\cos x = -1$

8. $\sin x = -\sqrt{2}/2$

9. $\cos x = \sqrt{3}/2$

10. $\sin x = 1/2$

11. $\sin x = 0$

12. $\cos x = 0$

13. $4\sin^2 x = 1$

14. $4\cos^2 x = 3$

15. $3\tan^2 x = 1$

Find all solutions x, $0 \leq x < 2\pi$, for the equations (without using a calculator):

16. $\cos 3x = -1/2$

17. $\sin 3x = -1/2$

18. $\tan 4x = 0$

19. $\tan 4x = 1$

20. $\sin(x/4) = \sqrt{3}/2$

21. $\cos(x/4) = \sqrt{3}/2$

22. $\tan 2x = -\sqrt{3}/3$

23. $\tan 2x = \sqrt{3}/3$

24. $\tan(2x/3) = 1$

25. $\tan(3x/4) = 1$

26. $\sin(2x/3) = 1/2$

27. $\cos(2x/3) = -1/2$

28. $3 \tan^2 3x = 1$

29. $4 \sin^2 3x = 1$

30. $4 \cos^2(x/4) = 3$

Find all solutions x in the interval $[0,2\pi)$ (without using a calculator):

31. $\sin^2 x \cos x = 0$

32. $2 \sin^2 x + \sin x = 1$

33. $\sin^2 x + 2\sin x + 1 = 0$

34. $\cos^2 x + 2 \cos x - 3 = 0$

35. $\csc^2 x - 4 = 0$

36. $2 \sin^2 x - \sin x - 3 = 0$

37. $4 \sin^3 x - \sin x = 0$

38. $2 \sin^4 x + \sin^2 x - 1 = 0$

39. $2 \cos^2 x + 3 \cos x + 1 = 0$

40. $2 \tan^2 x \cos x = \tan^2 x$

41. $2 \tan x \cos x - 2 \cos x + 1 = \tan x$

42. $\sqrt{3} \sin x \cos x + 2 \cos x = 2 + \sqrt{3} \sin x$

43. $\sin^2 2x = \sin 2x$

44. $4 \tan^2 3x = 3 \tan 3x$

Find all solutions x in the interval $[0,2\pi)$:

45. $\sin x + \cos x = 1$

46. $\sin x - \cos x = 1$

47. $2 \cos^2 x \sin x = \sin x$

48. $2(\sin 3x \cos x + \sin x \cos 3x) = \sqrt{3}$

49. $2 \sin x \cos 3x - 2 \sin 3x \cos x = \sqrt{3}$

50. $\sin 2x + \cos x = 0$

51. $\sin 4x = 2 \sin 2x$

52. $\sin x = \cos 2x$

53. $2(\cos^2 x - \sin^2 x) = 0$

54. $2(\sin^2 x - \cos^2 x) = \sqrt{3}$

55. $2(\cos 2x \cos x - \sin x \sin 2x) = 1$

56. $\sin x = \cos x$

57. $\tan x = \cot x$

58. $\sin 4x + 2 \sin 2x = 0$

59. $\sin(\pi - x) + \cos(\pi - x) = 1$

60. $\sin(\pi - x) + \cos(\pi/2 - x) = 1$

61. $2 \tan x + \sec^2 x = 0$

62. $\tan x = \sec x - 1$

63. $\cos 2x = \cos x - 1$

64. $\cos x - \sqrt{3} \sin x + 2 = 0$

65. $4 \sin 2x = \sin 4x$

66. $\cos x = \sin 2x$

67. $\sin 2x - \sin x + 2 \cos x - 1 = 0$

68. $2(\cos x + \sin x)^2 = 1$

69. $\sec x = 2$

70. $3 \sec^2 2x = 4$

71. $\csc x = -2$

72. $3 \csc^2 3x = 4$

73. $\sec^2 4x = 2$

✓ Check Yourself

Equations 1 – 12 are simple trigonometric equations.

1. The cosine is positive in quadrants I and IV. $x = \pi/4$ is the first quadrant solution and $x' = -\pi/4$ is the solution in quadrant IV. The cosine is periodic with period 2π. Therefore the set of all solutions is

    $x_n = \pi/4 + 2n\pi$, $x'_n = -\pi/4 + 2n\pi$, $n = 0, \pm 1, \pm 2, \ldots$

2. The tangent is positive in quadrants I and III. Then $x = \pi/4$ is the first quadrant solution and $x' = 5\pi/4$ is the solution in quadrant III. The tangent is periodic with period π. Therefore the set of all solutions is

    $x_n = \pi/4 + n\pi$, $n = 0, \pm 1, \pm 2, \ldots$

    Note that $x'_n = 5\pi/4 + n\pi = \pi/4 + (n+1)\pi = x_{n+1}$ so all of the solutions are included in the x_n formula.

3. The sine is negative in quadrants III and IV. Then $x = 7\pi/6$ is the third quadrant solution and $x' = -\pi/6$ is the solution in quadrant IV. The sine is periodic with period 2π. Therefore the set of all solutions is

    $x_n = 7\pi/6 + 2n\pi$, $x'_n = -\pi/6 + 2n\pi$, $n = 0, \pm 1, \pm 2, \ldots$

4. The cosine is negative in quadrants II and III. Then $x = 2\pi/3$ is the second quadrant solution and $x' = 4\pi/3$ is the solution in quadrant III. The cosine is periodic with period 2π. Therefore the set of all solutions is given by

    $x_n = 2\pi/3 + 2n\pi$, $x'_n = 4\pi/3 + 2n\pi$, $n = 0, +1, +2, \ldots$

5. Since $\sin(3\pi/2) = -1$ and the sine is 2π – periodic, the set of all solutions is given by $x_n = 3\pi/2 + 2n\pi$, $n = 0, \pm 1, \pm 2, \ldots$

6. Since $\tan 0 = 0$ and the tangent is π – periodic, the set of all solutions is given by
    $x_n = n\pi$, $n = 0, \pm 1, \pm 2, \ldots$

7. $x = \pi$ is the only x in $[0, 2\pi)$ with $\cos x = -1$. Then since cosine is 2π – periodic, the solution set consists of $x_n = \pi + 2n\pi$, $n = 0, \pm 1, \ldots$

8. This equation has solutions $x = 5\pi/4$ in quadrant III and $x' = -\pi/4$ in quadrant IV. Thus the solution set consists of the values

    $x_n = 5\pi/4 + 2n\pi$, $x'_n = -\pi/4 + 2n\pi$, $n = 0, +1, +2, \ldots$

9. The cosine is positive in quadrants I and IV. In particular, $\cos(\pi/6) = \cos(-\pi/6) = \sqrt{3}/2$. Since cosine is 2π – periodic, the set of solutions is

    $x_n = \pi/6 + 2n\pi$, $x'_n = -\pi/6 + 2n\pi$, $n = 0, +1, +2, \ldots$

10. The sine is positive in quadrants I and II. In particular, $\sin(\pi/6) = \sin(5\pi/6) = 1/2$. Since sine is 2π – periodic, the set of solutions is

    $x_n = \pi/6 + 2n\pi$, $x'_n = 5\pi/6 + 2n\pi$, $n = 0, \pm1, \pm2, \ldots$

11. Since $\sin 0 = \sin \pi = 0$, the sine equals zero at the points

    $x_n = 2n\pi$, $x'_n = \pi + 2n\pi$, $n = 0, \pm1, \pm2, \ldots$;

    That is, $\sin z_n = 0$ for $z_n = n\pi$, $n = 0, \pm1, \pm2, \ldots$

12. Since $\cos \pi/2 = \cos 3\pi/2 = 0$, the cosine equals zero at the points

    $x_n = \pi/2 + 2n\pi$, $x'_n = 3\pi/2 + 2n\pi$, $n = 0, \pm1, \pm2, \ldots$;

    That is, $\cos z_n = 0$ for $z_n = (2n+1)\pi/2$, $n = 0, \pm1, \pm2, \ldots$

Equations 13 – 15 are solved by factoring.

13. Write $4 \sin^2x - 1 = (2 \sin x - 1)(2 \sin x + 1) = 0$. For the product to equal zero, at least one of the factors must equal zero. Therefore

    $(2 \sin x - 1) = 0$ or $(2 \sin x + 1) = 0$.

    The solutions of these two equations are (see problems 3 and 10)

    $x_n = \pi/6 + 2n\pi$, $x'_n = 5\pi/6 + 2n\pi$, $n = 0, \pm1, \pm2, \ldots$

    and

    $x_n = 7\pi/6 + 2n\pi$, $x'_n = -\pi/6 + 2n\pi$, $n = 0, \pm1, \pm2, \ldots$

14. If $4 \cos^2x - 3 = (2 \cos x - \sqrt{3})(2 \cos x + \sqrt{3}) = 0$, then at least one of the factors must equal zero. That is, $2 \cos x = \sqrt{3}$ or $2 \cos x = -\sqrt{3}$.

    The solutions of these equations are (see problem 9)

    $x_n = \pi/6 + 2n\pi$, $x'_n = -\pi/6 + 2n\pi$, $n = 0, \pm1, \pm2, \ldots$

    or

    $x_n = 5\pi/6 + 2n\pi$, $x'_n = 7\pi/6 + 2n\pi$, $n = 0, \pm1, \pm2, \ldots$

15. If $3 \tan^2x - 1 = (\sqrt{3} \tan x - 1)(\sqrt{3} \tan x + 1) = 0$, then at least one of the factors must equal zero. Then

    $\sqrt{3} \tan x - 1 = 0$ or $\sqrt{3} \tan x + 1 = 0$

    The solutions of these equations are

    $x_n = \pi/6 + 2n\pi$, $x'_n = 7\pi/6 + 2n\pi$, $n = 0, \pm1, \pm2, \ldots$

    or

    $x_n = 5\pi/6 + 2n\pi$, $x'_n = -\pi/6 + 2n\pi$, $n = 0, \pm1, \pm2, \ldots$

Equations 16 – 27 involve multiple angles.

16. Let $z = 3x$. If the solutions of $\cos z = -1/2$ are

    $z_n = 2\pi/3 + 2n\pi$, $z'_n = 4\pi/3 + 2n\pi$, $n = 0, \pm1, \pm2, \ldots$

    then the full set of solutions for $\cos 3x = -1/2$ is

    $x_n = z_n/3 = 2\pi/9 + 2n\pi/3$, $n = 0, \pm1, \pm2, \ldots$

    $x'_n = z'_n/3 = 4\pi/9 + 2n\pi/3$, $n = 0, \pm1, \pm2, \ldots$

    The solutions in $[0, 2\pi)$ are then $x = 2\pi/9, 4\pi/9, 8\pi/9, 10\pi/9, 14\pi/9, 16\pi/9$.

17. Let $z = 3x$. If the solutions of $\sin z = -1/2$ are

    $z_n = 7\pi/6 + 2n\pi$, $z'_n = -\pi/6 + 2n\pi$, $n = 0, \pm 1, \pm 2, \ldots$

    then the full set of solutions for $\sin 3x = -1/2$ is

    $x_n = z_n/3 = 7\pi/18 + 2n\pi/3$, $n = 0, \pm 1, \pm 2, \ldots$

    $x'_n = z'_n/3 = -\pi/18 + 2n\pi/3$, $n = 0, \pm 1, \pm 2, \ldots$

    The solutions in $[0, 2\pi)$ are then $x = 7\pi/8, 11\pi/18, 19\pi/18, 23\pi/18, 31\pi/18, 35\pi/18$.

18. Let $z = 4x$. If the solutions of $\tan z = 0$ are $z_n = n\pi$, $n = 0, \pm 1, \ldots$ then the full set of solutions for $\tan 4x = 0$ consists of

    $x_n = n\pi/4$, $n = 0, \pm 1, \pm 2, \ldots$

    The solutions in $[0, 2\pi)$ are $x = n\pi/4$, $n = 0, 1, \ldots, 7$

19. Let $z = 4x$. If the solutions of $\tan z = 1$ are $z_n = \pi/4 + n\pi$, $n = 0, \pm 1, \ldots$ then the full set of solutions for $\tan 4x = 1$ consists of

    $x_n = z_n/4 = \pi/16 + n\pi/4$, $n = 0, \pm 1, \pm 2, \ldots$

    The solutions in $[0, 2\pi)$ are then $x = \pi/16 + n\pi/4$, $n = 0, 1, \ldots, 7$

20. Let $z = x/4$. If the solutions of $\sin z = \sqrt{3}/2$ are

    $z_n = \pi/3 + 2n\pi$, $z'_n = 2\pi/3 + 2n\pi$, $n = 0, \pm 1, \pm 2, \ldots$

    and the solutions of $\sin x/4 = \sqrt{3}/2$ are

    $x_n = 4z_n = 4\pi/3 + 8n\pi$, $x'_n = 4w_n = 8\pi/3 + 8n\pi$, $n = 0, \pm 1, \pm 2, \ldots$

    then the only solution in $[0, 2\pi)$ is $x = 4\pi/3$.

21. Let $z = x/4$. If the solutions of $\cos z = \sqrt{3}/2$ are

    $z_n = \pi/6 + 2n\pi$, $z'_n = -\pi/6 + 2n\pi$, $n = 0, \pm 1, \pm 2, \ldots$

    and the solutions of $\cos x/4 = \sqrt{3}/2$ are

    $x_n = 4z_n = 2\pi/3 + 8n\pi$, $x'_n = 4w_n = -2\pi/3 + 8n\pi$, $n = 0, \pm 1, \pm 2, \ldots$

    then the only solution in $[0, 2\pi)$ is $x = 2\pi/3$.

22. Let $z = 2x$. If the solutions of $\tan z = -\sqrt{3}/3$ are

    $z_n = 5\pi/6 + n\pi$, $n = 0, \pm 1, \pm 2, \ldots$

    and the solutions of $\tan 2x = -\sqrt{3}/3$ are

    $x_n = z_n/2 = 5\pi/12 + n\pi/2$, $n = 0, \pm 1, \pm 2, \ldots$

    then the solutions in $[0, 2\pi)$ are $x = 5\pi/12, 11\pi/12, 17\pi/12, 23\pi/12$.

23. Let $z = 2x$. If the solutions of $\tan z = \sqrt{3}/3$ are

    $z_n = \pi/6 + n\pi$, $n = 0, \pm 1, \pm 2, \ldots$

    and the solutions of $\tan 2x = \sqrt{3}/3$ are

    $x_n = z_n/2 = \pi/12 + n\pi/2$, $n = 0, \pm 1, \pm 2, \ldots$

    then the solutions in $[0, 2\pi)$ are $x = \pi/12, 7\pi/12, 13\pi/12, 19\pi/12$.

24. Let $z = 2x/3$. If the solutions of $\tan z = 1$ are

    $z_n = \pi/4 + n\pi$, $n = 0, \pm1, \pm2, \ldots$

    and the solutions of $\tan 2x/3 = 1$ are

    $x_n = 3z_n/2 = 3\pi/8 + 3n\pi/2$, $n = 0, \pm1, \pm2, \ldots$

    then the solutions in $[0, 2\pi)$ are $x = 3\pi/8, 15\pi/8$.

25. Let $z = 3x/4$. If the solutions of $\tan z = 1$ are

    $z_n = \pi/4 + n\pi$, $n = 0, \pm1, \pm2, \ldots$

    and the solutions of $\tan 3x/4 = 1$ are

    $x_n = 4z_n/3 = \pi/3 + 4n\pi/3$, $n = 0, \pm1, \pm2, \ldots$

    then the solutions in $[0, 2\pi)$ are $x = \pi/3, 5\pi/3$.

26. Let $z = 2x/3$. If the solutions of $\sin z = 1/2$ are

    $z_n = \pi/6 + 2n\pi$, $z'_n = 5\pi/6 + 2n\pi$, $n = 0, \pm1, \pm2, \ldots$

    and the solutions of $\sin 2x/3 = 1/2$ are

    $x_n = 3z_n/2 = \pi/4 + 3n\pi$, $x'_n = 3z'_n/2 = 5\pi/4 + 3n\pi$, $n = 0, \pm1, \ldots$

    then the solutions in $[0, 2\pi)$ are $x = \pi/4, 5\pi/4$.

27. Let $z = 2x/3$. If the solutions of $\cos z = -1/2$ are

    $z_n = 2\pi/3 + 2n\pi$, $z'_n = 4\pi/3 + 2n\pi$, $n = 0, \pm1, \pm2, \ldots$

    and the solutions of $\cos 2x/3 = -1/2$ are

    $x_n = 3z_n/2 = \pi + 3n\pi$, $x'_n = 3z'_n/2 = 2\pi + 3n\pi$, $n = 0, \pm1, \ldots$

    then the only solution in $[0, 2\pi)$ is $x = \pi$.

Equations 28 – 45 are solved by factoring

28. Let $z = 3x$. If the solutions of

    $3\tan^2 z - 1 = (\sqrt{3}\tan z - 1)(\sqrt{3}\tan z + 1) = 0$

    are (see problem 15)

    $z_n = \pi/6 + 2n\pi, 5\pi/6 + 2n\pi,$

    $7\pi/6 + 2n\pi, -\pi/6 + 2n\pi$, $n = 0, \pm1, \pm2, \ldots$

    then the solutions of $3\tan^2 3x = 1$ are

    $x_n = z_n/3 = \pi/18 + 2n\pi/3, 5\pi/18 + 2n\pi/3,$

    $7\pi/18 + 2n\pi/3, -\pi/18 + 2n\pi/3$, $n = 0, \pm1, \pm2, \ldots$

    The solutions in $[0, 2\pi)$ are $x = m\pi/18$ for $m = 1, 5, 7, 11, 13, 17, 19, 23, 25, 29, 31, 35$.

29. Let $z = 3x$. If the solutions of

    $4\sin^2 z - 1 = (2\sin z - 1)(2\sin z + 1) = 0$

    are (see problem 13)

    $z_n = -\pi/6 + 2n\pi, \pi/6 + 2n\pi,$

    $5\pi/6 + 2n\pi, 7\pi/6 + 2n\pi$, $n = 0, \pm1, \pm2, \ldots$

then the solutions of $4 \sin^2 3x = 1$ are

$x_n = z_n/3 = -\pi/18 + 2n\pi/3, \pi/18 + 2n\pi/3,$

$5\pi/18 + 2n\pi/3, 7\pi/18 + 2n\pi/3, n = 0, \pm1, \pm2, \ldots$

The solutions in $[0,2\pi)$ are $x = m\pi/18$ for $m = 1, 5, 7, 11, 13, 17, 19, 23, 25, 29, 31, 35.$

30. Let $z = x/4$. If the solutions of

$4 \cos^2 z - 3 = (2 \cos z - \sqrt{3})(2 \cos z + \sqrt{3}) = 0$

are (see problem 14)

$z_n = -\pi/6 + 2n\pi, \pi/6 + 2n\pi,$

$5\pi/6 + 2n\pi, 7\pi/6 + 2n\pi, n = 0, \pm1, \pm2, \ldots$

then the solutions of $4 \cos^2 (x/4) = 3$ are

$x_n = 4z_n = -2\pi/3 + 8n\pi, 2\pi/3 + 8n\pi,$

$10\pi/3 + 8n\pi, 14\pi/3 + 8n\pi, n = 0, \pm1, \pm2, \ldots$

The only solution in $[0,2\pi)$ is $x = 2\pi/3.$

31. $\sin^2 x \cos x = 0$ if $\sin^2 x = 0$ or $\cos x = 0.$

The solutions to $\sin^2 x = 0$ in $[0,2\pi)$ are $x = 0,\pi.$

The solutions to $\cos x = 0$ in $[0,2\pi)$ are $x = \pi/2, 3\pi/2.$

The solution set consists of the values $0, \pi/2, \pi, 3\pi/2.$

32. $2 \sin^2 x + \sin x - 1 = (2\sin x - 1)(\sin x + 1) = 0$ if

$(2\sin x - 1) = 0$ or $(\sin x + 1) = 0$

The solutions in $[0,2\pi)$ to $2\sin x = 1$ are $x = \pi/6, 5\pi/6.$

The solution in $[0,2\pi)$ to $\sin x = -1$ is $x = 3\pi/2.$

The solution set consists of the values $\pi/6, 5\pi/6, 3\pi/2.$

33. $\sin^2 x + 2\sin x + 1 = (\sin x + 1)^2 = 0$ if $\sin x = -1.$

The solution in $[0,2\pi)$ to $\sin x = -1$ is $x = 3\pi/2.$

34. $\cos^2 x + 2 \cos x - 3 = (\cos x - 1)(\cos x + 3) = 0$

The equation $\cos x = -3$ has no solutions since $-1 \le \cos x \le 1$ for all x, and $\cos x = 1$ for $x = 0,2\pi.$

The only solution in $[0,2\pi)$ is $x = 0.$

35. $\csc^2 x - 4 = (\csc x - 2)(\csc x + 2) = 0$ if $\csc x = 2$ or $\csc x = -2.$

The solutions in $[0,2\pi)$ to $\csc x = 2$ are $x = \pi/6, 5\pi/6.$

The solutions in $[0,2\pi)$ to $\csc x = -2$ are $x = 7\pi/6, 11\pi/6.$

The solution set consists of $\pi/6, 5\pi/6, 7\pi/6, 11\pi/6.$

36. $2 \sin^2 x - \sin x - 3 = (2 \sin x - 3)(\sin x + 1) = 0$

There are no solutions to $2\sin x = 3.$

The solution in $[0,2\pi)$ to $\sin x = -1$ is $x = 3\pi/2.$

37. $4 \sin^3 x - \sin x = \sin x (4 \sin^2 x - 1) = 0$

The solutions in $[0,2\pi)$ to $\sin x = 0$ are $x = 0, \pi$.

The solutions in $[0,2\pi)$ to $4 \sin^2 x = 1$ are $x = \pi/6, 5\pi/6, 7\pi/6$ and $11\pi/6$.

38. $2 \sin^4 x + \sin^2 x - 1 = (2 \sin^2 x - 1)(\sin^2 x + 1) = 0$

The solutions in $[0,2\pi)$ to $2\sin^2 x = 1$ are $x = \pi/4, 3\pi/4, 5\pi/4$, and $7\pi/4$.

There are no solutions to $\sin^2 x = -1$.

39. $2 \cos^2 x + 3 \cos x + 1 = (2\cos x + 1)(\cos x + 1) = 0$

The solutions in $[0,2\pi)$ to $2\cos x = -1$ are $x = 2\pi/3, 4\pi/3$.

The solution in $[0,2\pi)$ to $\cos x = -1$ is $x = \pi$.

40. $2 \tan^2 x \cos x - \tan^2 x = \tan^2 x (2\cos x - 1) = 0$

The solutions in $[0,2\pi)$ to $\tan^2 x = 0$ are $x = 0, \pi$.

The solutions in $[0,2\pi)$ to $2\cos x = 1$ are $x = \pi/3, 5\pi/3$.

41. $2 \tan x \cos x - 2 \cos x + 1 - \tan x = (\tan x - 1)(2\cos x - 1) = 0$

The solutions in $[0,2\pi)$ to $\tan x = 1$ are $x = \pi/4, 5\pi/4$.

The solutions in $[0,2\pi)$ to $2\cos x = 1$ are $x = \pi/3, 5\pi/3$.

42. $\sqrt{3} \sin x \cos x + 2 \cos x - 2 - \sqrt{3} \sin x = (\sqrt{3} \sin x + 2)(\cos x - 1) = 0$

There are no solutions to $\sqrt{3} \sin x = -2$.

The only solution in $[0,2\pi)$ to $\cos x = 1$ is $x = 0$.

43. $\sin^2 2x - \sin 2x = \sin 2x (\sin 2x - 1) = 0$

The solutions in $[0,2\pi)$ to $\sin 2x = 1$ are $x = \pi/4, 5\pi/4$.

The solutions in $[0,2\pi)$ to $\sin 2x = 0$ are $x = 0, \pi/2, \pi, 3\pi/2$.

44. $4 \tan^2 3x - 3 \tan 3x = \tan 3x (4 \tan 3x - 3) = 0$

The solutions in $[0,2\pi)$ to $4\tan 3x = 3$ are $x = 12.29° + n(60°)$ for $n = 0,1,...,5$.

The solutions in $[0,2\pi)$ to $\tan 3x = 0$ are $x = 0, \pi/3, 2\pi/3, \pi, 4\pi/3$, and $5\pi/3$.

Equations 45 – 75 are solved using identities

45. If $\sin x + \cos x = 1$, then squaring both sides leads to

$(\sin x + \cos x)^2 = \sin^2 x + 2 \sin x \cos x + \cos^2 x = 1$

$= 1 + 2 \sin x \cos x = 1$.

Thus, $\sin x \cos x = 0$, and

$\sin x = 0$ if $x = 0, \pi$ and $\cos x = 0$ if $x = \pi/2, 3\pi/2$.

Checking solutions in the original equation:

$x = 0$	$\sin 0 + \cos 0 = 1$	(solution checks)
$x = \pi/2$	$\sin(\pi/2) + \cos(\pi/2) = 1$	(solution checks)
$x = \pi$	$\sin \pi + \cos \pi = -1$	(extraneous solution)
$x = 3\pi/2$	$\sin(3\pi/2) + \cos(3\pi/2) = -1$	(extraneous solution)

46. If $\sin x - \cos x = 1$, then squaring both sides leads to

    $(\sin x - \cos x)^2 = \sin^2 x - 2 \sin x \cos x + \cos^2 x = 1$

    $= 1 - 2 \sin x \cos x = 1$.

    Thus, $\sin x \cos x = 0$, and

    $\sin x = 0$ if $x = 0, \pi$ and $\cos x = 0$ if $x = \pi/2, 3\pi/2$.

    Checking solutions in the original equation:

$x = 0$	$\sin 0 - \cos 0 = -1$	(extraneous solution)
$x = \pi/2$	$\sin(\pi/2) - \cos(\pi/2) = 1$	(solution checks)
$x = \pi$	$\sin \pi - \cos \pi = 1$	(solution checks)
$x = 3\pi/2$	$\sin(3\pi/2) + \cos(3\pi/2) = -1$	(extraneous solution)

47. $2 \cos^2 x \sin x - \sin x = \sin x (2 \cos^2 x - 1) = \sin x \cos 2x = 0$

    Thus, $\sin x = 0$ if $x = 0, \pi$

    and $\cos 2x = 0$ if $x = \pi/4, 3\pi/4, 5\pi/4$ and $7\pi/4$.

    Since the equation was only factored, and not raised to a power, all solutions satisfy the original equation.

48. The equation is equivalent to

    $(\sin 3x \cos x + \sin x \cos 3x) = \sin 4x = \sqrt{3}/2$.

    The solutions are $4x = m(\pi/3)$ for $m = 1, 2, 7, 8, 13, 14, 19, 20$.

    That is, the solutions with x in $[0, 2\pi)$ are $x = m(\pi/12)$ for $m = 1, 2, 7, 8, 13, 14, 19, 20$.

49. If $2\sin x \cos 3x - 2 \sin 3x \cos x = \sqrt{3}$, then

    $\sin x \cos 3x - \sin 3x \cos x = \sqrt{3}/2$

    $\sin (x - 3x) = \sqrt{3}/2$

    $\sin (-2x) = \sqrt{3}/2$

    $\sin 2x = -\sqrt{3}/2$

    and

    $2x = m(\pi/3)$ for $m = 4, 5, 10, 11$

    $x = m(\pi/6)$ for $m = 4, 5, 10, 11$ belong to $[0, 2\pi)$.

50. If $\sin 2x + \cos x = 2 \sin x \cos x + \cos x = (2 \sin x + 1) \cos x = 0$

    then $2\sin x + 1 = 0$ if $x = 7\pi/6, 11\pi/6$

    and $\cos x = 0$ if $x = \pi/2, 3\pi/2$.

    Since the equation was only factored, all solutions satisfy the original equation.

51. $\sin 4x - 2 \sin 2x = 2 \sin 2x \cos 2x - 2 \sin 2x$

    $$= 2 \sin 2x (\cos 2x - 1) = 0$$

    Thus, $\cos 2x - 1 = 0$ if $x = 0, \pi$

    and $\sin 2x = 0$ if $x = 0, \pi/2, \pi, 3\pi/2$.

    Since the equation was only factored, all solutions satisfy the original equation.

52. $\sin x - \cos 2x = \sin x - (1 - 2\sin^2 x) = 2\sin^2 x + \sin x - 1 = 0$

$$= (2\sin x - 1)(\sin x + 1) = 0$$

Thus, $2\sin x - 1 = 0$ if $x = \pi/6, 5\pi/6$

and $\sin x + 1 = 0$ if $x = 3\pi/2$.

Since the equation was only factored, all solutions satisfy the original equation.

53. $2(\cos^2 x - \sin^2 x) = 2\cos 2x = 0$ if $2x = 0, 2\pi$; i.e., if $x = 0, \pi$.

Since an identity was used to reduce the equation, all solutions satisfy the original equation.

54. $2(\sin^2 x - \cos^2 x) = -2\cos 2x = \sqrt{3}$ if $2x = m(\pi/6)$ for $m = 5, 7, 17, 19$.

That is, $x = m(\pi/12)$ for $m = 5, 7, 17, 19$.

Since an identity was used to reduce the equation, all solutions satisfy the original equation.

55. $2(\cos 2x \cos x - \sin x \sin 2x) = 2\cos 3x = 1$

Thus, $3x = m(\pi/3)$ for $m = 1, 5, 7, 11, 13, 17$

and $x = m(\pi/9)$ for $m = 1, 5, 7, 11, 13, 17$.

Since an identity was used to reduce the equation, all solutions satisfy the original equation.

56. $\sin x = \cos x$ if $\tan x = 1$; i.e. $x = \pi/4, 5\pi/4$. Both solutions check.

57. $\tan x = \cot x$ if $\tan^2 x = \tan x \cot x = 1$.

$\tan^2 x = 1$ if $x = \pi/4, 3\pi/4, 5\pi/4, 7\pi/4$.

All these solutions check.

58. $\sin 4x + 2\sin 2x = 2\sin 2x \cos 2x + 2\sin 2x$

$$= 2\sin 2x (\cos 2x + 1) = 0$$

Thus, $\cos 2x + 1 = 0$ if $x = \pi/2, 3\pi/2$

and $\sin 2x = 0$ if $x = 0, \pi/2, \pi, 3\pi/2$.

Since the equation was only factored, all solutions satisfy the original equation.

59. $\sin(\pi - x) + \cos(\pi - x) = 1$ if $(\pi - x) = 0, \pi/2$ (see problem 45)

Thus, $x = \pi/2, \pi$.

60. Since $\sin(\pi - x) = \sin x$, and $\cos(\pi/2 - x) = \sin x$

$\sin(\pi - x) + \cos(\pi/2 - x) = 2\sin x = 1$

if $x = \pi/6, 5\pi/6$.

61. $2\tan x + \sec^2 x = 2\tan x + (\tan^2 x + 1)$

$$= (\tan x + 1)^2 = 0$$

Thus, $\tan x + 1 = 0$ if $x = 3\pi/4, 7\pi/4$.

62. $\tan x + 1 = \sec x$ is equivalent to $\sin x + \cos x = 1$ (multiply the equation by $\cos x$) at all values for which it is defined.

Thus, $x = 0, \pi/2$. The solution $x = 0$ checks, but $\tan x$ is undefined at $x = \pi/2$, so this is an extraneous solution.

Alternate solution: $(\tan x + 1)^2 = \sec^2 x$ or

$\tan^2 x + 2 \tan x + 1 = \tan^2 x + 1$. Thus,

$\tan x = 0$ and $x = 0, \pi$. Again $x = \pi$ is extraneous.

63. $\cos 2x - \cos x + 1 = 2 \cos^2 x - 1 - \cos x + 1 = \cos x (2 \cos x - 1)$

Thus, $\cos x = 0$ if $x = \pi/2, 3\pi/2$

$2 \cos x = 1$ if $x = \pi/3, 5\pi/3$

Since an identity was used to reduce the equation, all solutions satisfy the original equation.

64. If $\cos x - \sqrt{3} \sin x + 2 = 0$

then $(1/2) \cos x - (\sqrt{3}/2) \sin x = \sin \pi/6 \cos x - \cos \pi/6 \sin x = \sin (\pi/6 - x) = -1$.

Thus, $\pi/6 - x = 3\pi/2 + 2n\pi$,

and $x = -4\pi/3 + 2n\pi$,

so $x = 2\pi/3$ is the only solution in $[0, 2\pi]$.

65. If $4 \sin 2x = \sin 4x$ then $4 \sin 2x = 2 \sin 2x \cos 2x$;

then $2 \sin 2x (2 - \cos 2x) = 0$

and $\sin 2x = 0$ if $x = 0, \pi/2, \pi, 3\pi/2$.

$\cos 2x = 2$ has no solutions.

66. $\cos x - \sin 2x = \cos x - 2 \sin x \cos x = \cos x (1 - 2 \sin x) = 0$

Then $\cos x = 0$ if $x = \pi/2, 3\pi/2$.

$2 \sin x = 1$ if $x = \pi/6, 5\pi/6$.

67. $\sin 2x - \sin x + 2 \cos x - 1 = 2 \sin x \cos x - \sin x + 2 \cos x - 1$

$$= (\sin x + 1)(2 \cos x - 1) = 0$$

Thus, $2 \cos x - 1 = 0$ if $x = \pi/3, 5\pi/3$.

$\sin x + 1 = 0$ if $x = 3\pi/2$.

68. $(\cos x + \sin x)^2 = \cos^2 x + 2 \cos x \sin x + \sin^2 x = 1/2$

$\cos^2 x + \sin^2 x + 2 \sin x \cos x = 1 + 2 \sin x \cos x = 1/2$

$\sin 2x = -1/2$

Thus, $\sin 2x = -1/2$ if $x = m\pi/12$ for $m = 7, 11, 19, 23$.

All solutions check.

69. $\sec x = 2$ if $\cos x = 1/2$. Then $x = \pi/3, 5\pi/3$.

70. $3 \sec^2 2x = 4$ is equivalent to $\sec 2x = \pm 2/\sqrt{3}$.

Thus, $\cos 2x = \pm(\sqrt{3}/2)$

and $2x = m\pi/6$ for

$m = 1, 5, 7, 11, 13, 17, 19, 23$.

The solutions x in $[0, 2\pi)$ are $x = m\pi/12$ for $m = 1, 5, 7, 11, 13, 17, 19,$ and 23.

71.   csc $x = -2$ if sin $x = -1/2$. Then $x = 7\pi/6, 11\pi/6$

72. $3 \csc^2 3x = 4$ is equivalent to $3\csc^2 3x - 3 = 1$;

Thus, $\cot^2 3x = 1/3$.

Then, $3x = m\pi/3$ for $m = 1, 2, 4, 5, 7, 8, 10, 11, 13, 14, 16, 17$.

The solutions x in $[0, 2\pi)$ are

$x = m\pi/9$ for $m = 1, 2, 4, 5, 7, 8, 10, 11, 13, 14, 16, 17$.

73. $\sec^2 4x = 2$ if cos $4x = \pm 1/\sqrt{2}$.

Thus, $4x = m\pi/4$ for $m = 1, 3, 5, 7, 9, 11, 13, 15, 17, 19, 21, 23, 25, 27, 29, 31$.

The solutions x in $[0, 2\pi)$ are then

$x = m\pi/16$ for $m = 1, 3, 5, 7, 9, 11, 13, 15, 17, 19, 21, 23, 25, 27, 29, 31$.

Grade Yourself

Circle the numbers of the questions you missed, then fill in the total incorrect for each topic. If you answered more than three questions incorrectly, you need to focus on that topic. (If a topic has less than three questions and you had at least one wrong, we suggest you study that topic also. Read your textbook, a review book, or ask your teacher for help.)

Subject: Trigonometric Equations

Topic	Question Numbers	Number Incorrect
Simple trigonometric equations	1, 2, 3, 4, 5, 6, 7, 8, 9, 10, 11, 12	
Factoring	13, 14, 15, 28, 29, 30, 31, 32, 33, 34, 35, 36, 37, 38, 39, 40, 41, 42, 43, 44, 45, 46	
Multiple angles	16, 17, 18, 19, 20, 21, 22, 23, 24, 25, 26, 27	
Identities	47, 48, 49, 50, 51, 52, 53, 54, 55, 56, 57, 58, 59, 60, 61, 62, 63, 64, 65, 66, 67, 68, 69, 70, 71, 72, 73	

Trigonometry Tables

8

The following tables will help you in your calculations. Included are:

A. Common Logarithms of Numbers

B. Values of Trigonometric Functions

C. Logarithms of Trigonometric Functions

Table A: Common Logarithms of Numbers*

N	0	1	2	3	4	5	6	7	8	9
55	7404	7412	7419	7427	7435	7443	7451	7459	7466	7474
56	7482	7490	7497	7505	7513	7520	7528	7536	7543	7551
57	7559	7566	7574	7582	7589	7597	7604	7612	7619	7627
58	7634	7642	7649	7657	7664	7672	7679	7686	7694	7701
59	7709	7716	7723	7731	7738	7745	7752	7760	7767	7774
60	7782	7789	7796	7803	7810	7818	7825	7832	7839	7846
61	7853	7860	7868	7875	7882	7889	7896	7903	7910	7917
62	7924	7931	7938	7945	7952	7959	7966	7973	7980	7987
63	7993	8000	8007	8014	8021	8028	8035	8041	8048	8055
64	8062	8069	8075	8082	8089	8096	8102	8109	8116	8122
65	8129	8136	8142	8149	8156	8162	8169	8176	8182	8189
66	8195	8202	8209	8215	8222	8228	8235	8241	8248	8254
67	8261	8267	8274	8280	8287	8293	8299	8306	8312	8319
68	8325	8331	8338	8344	8351	8357	8363	8370	8376	8382
69	8388	8395	8401	8407	8414	8420	8426	8432	8439	8445
70	8451	8457	8463	8470	8476	8482	8488	8494	8500	8506
71	8513	8519	8525	8531	8537	8543	8549	8555	8561	8567
72	8573	8579	8585	8591	8597	8603	8609	8615	8621	8627
73	8633	8639	8645	8651	8657	8663	8669	8675	8681	8686
74	8692	8698	8704	8710	8716	8722	8727	8733	8739	8745
75	8751	8756	8762	8768	8774	8779	8785	8791	8797	8802
76	8808	8814	8820	8825	8831	8837	8842	8848	8854	8859
77	8865	8871	8876	8882	8887	8893	8899	8904	8910	8915
78	8921	8927	8932	8938	8943	8949	8954	8960	8965	8971
79	8976	8982	8987	8993	8998	9004	9009	9015	9020	9025
80	9031	9036	9042	9047	9053	9058	9063	9069	9074	9079
81	9085	9090	9096	9101	9106	9112	9117	9122	9128	9133
82	9138	9143	9149	9154	9159	9165	9170	9175	9180	9186
83	9191	9196	9201	9206	9212	9217	9222	9227	9232	9238
84	9243	9248	9253	9258	9263	9269	9274	9279	9284	9289
85	9294	9299	9304	9309	9315	9320	9325	9330	9335	9340
86	9345	9350	9355	9360	9365	9370	9375	9380	9385	9390
87	9395	9400	9405	9410	9415	9420	9425	9430	9435	9440
88	9445	9450	9455	9460	9465	9469	9474	9479	9484	9489
89	9494	9499	9504	9509	9513	9518	9523	9528	9533	9538
90	9542	9547	9552	9557	9562	9566	9571	9576	9581	9586
91	9590	9595	9600	9605	9609	9614	9619	9624	9628	9633
92	9638	9643	9647	9652	9657	9661	9666	9671	9675	9680
93	9685	9689	9694	9699	9703	9708	9713	9717	9722	9727
94	9731	9736	9741	9745	9750	9754	9759	9763	9768	9773
95	9777	9782	9786	9791	9795	9800	9805	9809	9814	9818
96	9823	9827	9832	9836	9841	9845	9850	9854	9859	9863
97	9868	9872	9877	9881	9886	9890	9894	9899	9903	9908
98	9912	9917	9921	9926	9930	9934	9939	9943	9948	9952
99	9956	9961	9965	9969	9974	9978	9983	9987	9991	9996
N	0	1	2	3	4	5	6	7	8	9

* This table gives the mantissas of numbers with the decimal point omitted in each case. Characteristics are determined from the numbers by inspection.

Table B: Values of Trigonometric Functions

Angle	Sin	Cos	Tan	Cot		
24° 00′	.4067	.9135	.4452	2.2460	66°	00′
10	.4094	.9124	.4487	2.2286		50
20	.4120	.9112	.4522	2.2113		40
30	.4147	.9100	.4557	2.1943		30
40	.4173	.9088	.4592	2.1775		20
50	.4200	.9075	.4628	2.1609		10
25° 00′	.4226	.9063	.4663	2.1445	65°	00′
10	.4253	.9051	.4699	2.1283		50
20	.4279	.9038	.4734	2.1123		40
30	.4305	.9026	.4770	2.0965		30
40	.4331	.9013	.4806	2.0809		20
50	.4358	.9001	.4841	2.0655		10
26° 00′	.4384	.8988	.4877	2.0503	64°	00′
10	.4410	.8975	.4913	2.0353		50
20	.4436	.8962	.4950	2.0204		40
30	.4462	.8949	.4986	2.0057		30
40	.4488	.8936	.5022	1.9912		20
50	.4514	.8923	.5059	1.9768		10
27° 00′	.4540	.8910	.5095	1.9626	63°	00′
10	.4566	.8897	.5132	1.9486		50
20	.4592	.8884	.5169	1.9347		40
30	.4617	.8870	.5206	1.9210		30
40	.4643	.8857	.5243	1.9074		20
50	.4669	.8843	.5280	1.8940		10
28° 00′	.4695	.8829	.5317	1.8807	62°	00′
10	.4720	.8816	.5354	1.8676		50
20	.4746	.8802	.5392	1.8546		40
30	.4772	.8788	.5430	1.8418		30
40	.4797	.8774	.5467	1.8291		20
50	.4823	.8760	.5505	1.8165		10
29° 00′	.4848	.8746	.5543	1.8040	61°	00′
10	.4874	.8732	.5581	1.7917		50
20	.4899	.8718	.5619	1.7796		40
30	.4924	.8704	.5658	1.7675		30
40	.4950	.8689	.5696	1.7556		20
50	.4975	.8675	.5735	1.7437		10
30° 00′	.5000	.8660	.5774	1.7321	60°	00′
10	.5025	.8646	.5812	1.7205		50
20	.5050	.8631	.5851	1.7090		40
30	.5075	.8616	.5890	1.6977		30
40	.5100	.8601	.5930	1.6864		20
50	.5125	.8587	.5969	1.6753		10
31° 00′	.5150	.8572	.6009	1.6643	59°	00′
10	.5175	.8557	.6048	1.6534		50
20	.5200	.8542	.6088	1.6426		40
30	.5225	.8526	.6128	1.6319		30
40	.5250	.8511	.6168	1.6212		20
50	.5275	.8496	.6208	1.6107		10
32° 00′	.5299	.8480	.6249	1.6003	58°	00′
10	.5324	.8465	.6289	1.5900		50
20	.5348	.8450	.6330	1.5798		40
30	.5373	.8434	.6371	1.5697		30
40	.5398	.8418	.6412	1.5597		20
50	.5422	.8403	.6453	1.5497		10
33° 00′	.5446	.8387	.6494	1.5399	57°	00′
10	.5471	.8371	.6536	1.5301		50
20	.5495	.8355	.6577	1.5204		40
30	.5519	.8339	.6619	1.5108		30
40	.5544	.8323	.6661	1.5013		20
50	.5568	.8307	.6703	1.4919		10
34° 00′	.5592	.8290	.6745	1.4826	56°	00′
10	.5616	.8274	.6787	1.4733		50
20	.5640	.8258	.6830	1.4641		40
30	.5664	.8241	.6873	1.4550		30
40	.5688	.8225	.6916	1.4460		20
50	.5712	.8208	.6959	1.4370		10
35° 00′	.5736	.8192	.7002	1.4281	55°	00′
10	.5760	.8175	.7046	1.4193		50
20	.5783	.8158	.7089	1.4106		40
30	.5807	.8141	.7133	1.4019		30
40	.5831	.8124	.7177	1.3934		20
50	.5854	.8107	.7221	1.3848		10
36° 00′	.5878	.8090	.7265	1.3764	54°	00′
	Cos	Sin	Cot	Tan	Angle	

Angle	Sin	Cos	Tan	Cot		
36° 00′	.5878	.8090	.7265	1.3764	54°	00′
10	.5901	.8073	.7310	1.3680		50
20	.5925	.8056	.7355	1.3597		40
30	.5948	.8039	.7400	1.3514		30
40	.5972	.8021	.7445	1.3432		20
50	.5995	.8004	.7490	1.3351		10
37° 00′	.6018	.7986	.7536	1.3270	53°	00′
10	.6041	.7969	.7581	1.3190		50
20	.6065	.7951	.7627	1.3111		40
30	.6088	.7934	.7673	1.3032		30
40	.6111	.7916	.7720	1.2954		20
50	.6134	.7898	.7766	1.2876		10
38° 00′	.6157	.7880	.7813	1.2799	52°	00′
10	.6180	.7862	.7860	1.2723		50
20	.6202	.7844	.7907	1.2647		40
30	.6225	.7826	.7954	1.2572		30
40	.6248	.7808	.8002	1.2497		20
50	.6271	.7790	.8050	1.2423		10
39° 00′	.6293	.7771	.8098	1.2349	51°	00′
10	.6316	.7753	.8146	1.2276		50
20	.6338	.7735	.8195	1.2203		40
30	.6361	.7716	.8243	1.2131		30
40	.6383	.7698	.8292	1.2059		20
50	.6406	.7679	.8342	1.1988		10
40° 00′	.6428	.7660	.8391	1.1918	50°	00′
10	.6450	.7642	.8441	1.1847		50
20	.6472	.7623	.8491	1.1778		40
30	.6494	.7604	.8541	1.1708		30
40	.6517	.7585	.8591	1.1640		20
50	.6539	.7566	.8642	1.1571		10
41° 00′	.6561	.7547	.8693	1.1504	49°	00′
10	.6583	.7528	.8744	1.1436		50
20	.6604	.7509	.8796	1.1369		40
30	.6626	.7490	.8847	1.1303		30
40	.6648	.7470	.8899	1.1237		20
50	.6670	.7451	.8952	1.1171		10
42° 00′	.6691	.7431	.9004	1.1106	48°	00′
10	.6713	.7412	.9057	1.1041		50
20	.6734	.7392	.9110	1.0977		40
30	.6756	.7373	.9163	1.0913		30
40	.6777	.7353	.9217	1.0850		20
50	.6799	.7333	.9271	1.0786		10
43° 00′	.6820	.7314	.9325	1.0724	47°	00′
10	.6841	.7294	.9380	1.0661		50
20	.6862	.7274	.9435	1.0599		40
30	.6884	.7254	.9490	1.0538		30
40	.6905	.7234	.9545	1.0477		20
50	.6926	.7214	.9601	1.0416		10
44° 00′	.6947	.7193	.9657	1.0355	46°	00′
10	.6967	.7173	.9713	1.0295		50
20	.6988	.7153	.9770	1.0235		40
30	.7009	.7133	.9827	1.0176		30
40	.7030	.7112	.9884	1.0117		20
50	.7050	.7092	.9942	1.0058		10
45° 00′	.7071	.7071	1.0000	1.0000	45°	00′
	Cos	Sin	Cot	Tan	Angle	

Table C: Logarithms of Trigonometric Functions*

Angle	L Sin	L Cos	L Tan	L Cot	
24° 00′	9.6093	9.9607	9.6486	10.3514	66° 00′
10	9.6121	9.9602	9.6520	10.3480	50
20	9.6149	9.9596	9.6553	10.3447	40
30	9.6177	9.9590	9.6587	10.3413	30
40	9.6205	9.9584	9.6620	10.3380	20
50	9.6232	9.9579	9.6654	10.3346	10
25° 00′	9.6259	9.9573	9.6687	10.3313	65° 00′
10	9.6286	9.9567	9.6720	10.3280	50
20	9.6313	9.9561	9.6752	10.3248	40
30	9.6340	9.9555	9.6785	10.3215	30
40	9.6366	9.9549	9.6817	10.3183	20
50	9.6392	9.9543	9.6850	10.3150	10
26° 00′	9.6418	9.9537	9.6882	10.3118	64° 00′
10	9.6444	9.9530	9.6914	10.3086	50
20	9.6470	9.9524	9.6946	10.3054	40
30	9.6495	9.9518	9.6977	10.3023	30
40	9.6521	9.9512	9.7009	10.2991	20
50	9.6546	9.9505	9.7040	10.2960	10
27° 00′	9.6570	9.9499	9.7072	10.2928	63° 00′
10	9.6595	9.9492	9.7103	10.2897	50
20	9.6620	9.9486	9.7134	10.2866	40
30	9.6644	9.9479	9.7165	10.2835	30
40	9.6668	9.9473	9.7196	10.2804	20
50	9.6692	9.9466	9.7226	10.2774	10
28° 00′	9.6716	9.9459	9.7257	10.2743	62° 00′
10	9.6740	9.9453	9.7287	10.2713	50
20	9.6763	9.9446	9.7317	10.2683	40
30	9.6787	9.9439	9.7348	10.2652	30
40	9.6810	9.9432	9.7378	10.2622	20
50	9.6833	9.9425	9.7408	10.2592	10
29° 00′	9.6856	9.9418	9.7438	10.2562	61° 00′
10	9.6878	9.9411	9.7467	10.2533	50
20	9.6901	9.9404	9.7497	10.2503	40
30	9.6923	9.9397	9.7526	10.2474	30
40	9.6946	9.9390	9.7556	10.2444	20
50	9.6968	9.9383	9.7585	10.2415	10
30° 00′	9.6990	9.9375	9.7614	10.2386	60° 00′
10	9.7012	9.9368	9.7644	10.2356	50
20	9.7033	9.9361	9.7673	10.2327	40
30	9.7055	9.9353	9.7701	10.2299	30
40	9.7076	9.9346	9.7730	10.2270	20
50	9.7097	9.9338	9.7759	10.2241	10
31° 00′	9.7118	9.9331	9.7788	10.2212	59° 00′
10	9.7139	9.9323	9.7816	10.2184	50
20	9.7160	9.9315	9.7845	10.2155	40
30	9.7181	9.9308	9.7873	10.2127	30
40	9.7201	9.9300	9.7902	10.2098	20
50	9.7222	9.9292	9.7930	10.2070	10
32° 00′	9.7242	9.9284	9.7958	10.2042	58° 00′
10	9.7262	9.9276	9.7986	10.2014	50
20	9.7282	9.9268	9.8014	10.1986	40
30	9.7302	9.9260	9.8042	10.1958	30
40	9.7322	9.9252	9.8070	10.1930	20
50	9.7342	9.9244	9.8097	10.1903	10
33° 00′	9.7361	9.9236	9.8125	10.1875	57° 00′
10	9.7380	9.9228	9.8153	10.1847	50
20	9.7400	9.9219	9.8180	10.1820	40
30	9.7419	9.9211	9.8208	10.1792	30
40	9.7438	9.9203	9.8235	10.1765	20
50	9.7457	9.9194	9.8263	10.1737	10
34° 00′	9.7476	9.9186	9.8290	10.1710	56° 00′
10	9.7494	9.9177	9.8317	10.1683	50
20	9.7513	9.9169	9.8344	10.1656	40
30	9.7531	9.9160	9.8371	10.1629	30
40	9.7550	9.9151	9.8398	10.1602	20
50	9.7568	9.9142	9.8425	10.1575	10
35° 00′	9.7586	9.9134	9.8452	10.1548	55° 00′
10	9.7604	9.9125	9.8479	10.1521	50
20	9.7622	9.9116	9.8506	10.1494	40
30	9.7640	9.9107	9.8533	10.1467	30
40	9.7657	9.9098	9.8559	10.1441	20
50	9.7675	9.9089	9.8586	10.1414	10
36° 00′	9.7692	9.9080	9.8613	10.1387	54° 00′
	L Cos	L Sin	L Cot	L Tan	Angle

Angle	L Sin	L Cos	L Tan	L Cot	
36° 00′	9.7692	9.9080	9.8613	10.1387	54° 00′
10	9.7710	9.9070	9.8639	10.1361	50
20	9.7727	9.9061	9.8666	10.1334	40
30	9.7744	9.9052	9.8692	10.1308	30
40	9.7761	9.9042	9.8718	10.1282	20
50	9.7778	9.9033	9.8745	10.1255	10
37° 00′	9.7795	9.9023	9.8771	10.1229	53° 00′
10	9.7811	9.9014	9.8797	10.1203	50
20	9.7828	9.9004	9.8824	10.1176	40
30	9.7844	9.8995	9.8850	10.1150	30
40	9.7861	9.8985	9.8876	10.1124	20
50	9.7877	9.8975	9.8902	10.1098	10
38° 00′	9.7893	9.8965	9.8928	10.1072	52° 00′
10	9.7910	9.8955	9.8954	10.1046	50
20	9.7926	9.8945	9.8980	10.1020	40
30	9.7941	9.8935	9.9006	10.0994	30
40	9.7957	9.8925	9.9032	10.0968	20
50	9.7973	9.8915	9.9058	10.0942	10
39° 00′	9.7989	9.8905	9.9084	10.0916	51° 00′
10	9.8004	9.8895	9.9110	10.0890	50
20	9.8020	9.8884	9.9135	10.0865	40
30	9.8035	9.8874	9.9161	10.0839	30
40	9.8050	9.8864	9.9187	10.0813	20
50	9.8066	9.8853	9.9212	10.0788	10
40° 00′	9.8081	9.8843	9.9238	10.0762	50° 00′
10	9.8096	9.8832	9.9264	10.0736	50
20	9.8111	9.8821	9.9289	10.0711	40
30	9.8125	9.8810	9.9315	10.0685	30
40	9.8140	9.8800	9.9341	10.0659	20
50	9.8155	9.8789	9.9366	10.0634	10
41° 00′	9.8169	9.8778	9.9392	10.0608	49° 00′
10	9.8184	9.8767	9.9417	10.0583	50
20	9.8198	9.8756	9.9443	10.0557	40
30	9.8213	9.8745	9.9468	10.0532	30
40	9.8227	9.8733	9.9494	10.0506	20
50	9.8241	9.8722	9.9519	10.0481	10
42° 00′	9.8255	9.8711	9.9544	10.0456	48° 00′
10	9.8269	9.8699	9.9570	10.0430	50
20	9.8283	9.8688	9.9595	10.0405	40
30	9.8297	9.8676	9.9621	10.0379	30
40	9.8311	9.8665	9.9646	10.0354	20
50	9.8324	9.8653	9.9671	10.0329	10
43° 00′	9.8338	9.8641	9.9697	10.0303	47° 00′
10	9.8351	9.8629	9.9722	10.0278	50
20	9.8365	9.8618	9.9747	10.0253	40
30	9.8378	9.8606	9.9772	10.0228	30
40	9.8391	9.8594	9.9798	10.0202	20
50	9.8405	9.8582	9.9823	10.0177	10
44° 00′	9.8418	9.8569	9.9848	10.0152	46° 00′
10	9.8431	9.8557	9.9874	10.0126	50
20	9.8444	9.8545	9.9899	10.0101	40
30	9.8457	9.8532	9.9924	10.0076	30
40	9.8469	9.8520	9.9949	10.0051	20
50	9.8482	9.8507	9.9975	10.0025	10
45° 00′	9.8495	9.8495	10.0000	10.0000	45° 00′
	L Cos	L Sin	L Cot	L Tan	Angle

* These tables give the logarithms increased by 10. Hence in each case 10 should be subtracted.